THE 100 GREATEST CORPORATE and INDUSTRIAL ADS

THE 100 GREATEST CORPORATE and INDUSTRIAL ADS

Fred C. Poppe

VNR VAN NOSTRAND REINHOLD COMPANY
NEW YORK CINCINNATI TORONTO LONDON MELBOURNE

Copyright © 1983 by Van Nostrand Reinhold Company Inc.

Library of Congress Catalog Card Number: 82-20047
ISBN: 0-442-27246-4

All rights reserved. No part of this work covered by the copyright hereon may be reproduced or used in any form or by any means—graphic, electronic, or mechanical, including photocopying, recording, taping, or information storage and retrieval systems—without permission of the publisher.

Manufactured in the United States of America

Published by Van Nostrand Reinhold Company Inc.
135 West 50th Street, New York, N.Y. 10020

Van Nostrand Reinhold Publishing
1410 Birchmount Road
Scarborough, Ontario MIP 2E7, Canada

Van Nostrand Reinhold
480 Latrobe Street
Melbourne, Victoria 3000, Australia

Van Nostrand Reinhold Company Limited
Molly Millars Lane
Wokingham, Berkshire, England

15 14 13 12 11 10 9 8 7 6 5 4 3 2 1

Library of Congress Cataloging in Publication Data
Main entry under title:

The 100 greatest corporate and industrial ads.

    Includes index.
    1. Advertising, Industrial.   I. Poppe, Fred C.
II. Title: One hundred greatest corporate and industrial ads.
HF5823.A12   1983   659.1   82-20047
ISBN 0-442-27246-4

TO INEZ, STEVE AND ELLEN
A super wife and two neat kids.

Foreword

So-Called Industrial Advertising

While the word *consumer*—as in consumer advertising—has the sound of something being enjoyably swallowed, the word *industrial*—as in industrial advertising—has, to my ears at least, the sound of a factory or plant. As a result, when I speak of industrial advertising I am impelled to refer to it as so-called industrial advertising; and this book (I think you as a reader of its pages will agree), will suggest to you why I am so inclined and impelled.

The term *industrial advertising* suggests a kind of advertising as heavy and ponderous as the kind of products and services written about. And no small amount of it *is* heavy and ponderous—just as no small amount of consumer advertising is inexplicably uninformative and written as if its primary purpose were to please management rather than to enlighten the public. I attribute this chiefly to a lack of sufficient talent to do well the enormous amount of copy that must be turned out today for advertisers of both industrial and consumer products, as well as the lack of a tradition of copy excellence in both fields. I welcome this book primarily because I believe it can contribute to the eventual creation of such a tradition (which is sorely needed in advertising generally).

Like trade advertising, most industrial advertising appears in relatively low-cost media. The excuse usually given for poorly done advertising in those media is that the agency commission on the space is inadequate to compensate exceptionally talented writers or art directors. This necessarily implies that most advertising agencies operate on the commission system, and that advertisers generally are strongly disinclined to pay a fee for superior performance. And, of course, it overlooks the deplorable fact that agencies put their highest paid creative people on accounts deep in highly profitable television despite survey after survey showing that television commercials are considered by the public to be the most irritating form of advertising. These surveys invariably reflect the most common criticism of television commercials being "an insult to my intelligence." So much for the relationship between money and quality. The latter ultimately produces the former, but the former does not necessarily produce the latter.

Despite the efforts of the general semanticists to make language as exact as mathematics, communication still remains an art; and since it is involved with communication, advertising writing is primarily an art. And the arts, including copywriting, benefit chiefly from devotion to a tradition of excellence. And intense devotion to such a tradition is least likely to come into being or remain intense through what it earns.

This book, however, can prove an important stimulus to such devotion. The examples it contains of outstanding and memorable advertisements done for industrial advertisers—some of them closer to corporate rather than so-called industrial advertising—provide examples of style and execution which a copywriter given an industrial writing assignment can wisely contemplate and study. Insofar as industrial advertisers are concerned, they could most profitably, in my opinion, see to it that any copywriter entrusted with their advertising has a copy of it constantly on his desk. In his office and at home.

WALTER WEIR
Temple University
School of Communications and Theater
Philadelphia, Pennsylvania

Preface

A little more than a year ago when I started this book, I was terrified by the awesome, interminable task of writing. My trepidation couldn't have been further off-target. The writing part was a cinch. It was the research, compilation and selection of the advertisements which caused me the most grief. It all started simply enough. Through the courtesy of the Business/Professional Advertising Association's managing director, Ron Coleman, I sent a letter to some 4,000 members, imploring them to send me ad nominations for inclusion in this opus. Within weeks I had hundreds of advertisements to cull and consider. I was in heaven (and up to my butt in ad reprints). My office looked like the Collier mansion, or, at the very least, Charles Manson's mansion. Unfortunately, many B/PAA members nominated the same ads. Out of all those nominations, I received 10 worthwhile ones. At least 50 B/PAAer's led me to the famous McGraw-Hill man-in-the-chair ad. Another 20 or so selected the current United Technologies *Wall Street Journal* corporate campaign. A dozen letters came in suggesting the Rockwell Reports. It seemed every time I queried experts they all remembered the same "greatest" ads. But luckily, I had many helpers out there pushing for me. People like Bill Cowan, for instance. Bill, a copy consultant from Waban, Massachusetts of all places (it's near Boston), touted me on to about 15 people who sent me to a bunch of their friends and associates. All of a sudden it started working in geometric progression. A lot of friends' friends were helpful, but the follow-up to letters and phone calls was still painstakingly slow. (Someday try to find a person who saved a reprint of a 1961 corporate ad that appeared in the now-defunct *Saturday Evening Post*). Many people moved, some changed jobs, and a few even had the audacity to die before I could get to them.

Finally it all came together. I got my magic 100. I'm happy with them and I hope you are too. And, I'll bet this book is going to generate a lot more activity. I know that I have missed at least another 100 "greatest" or so. They're out there somewhere. Probably some are better and more deserving to be in this book than those we have selected. That suits me fine. Like all copywriters and creative types, I'll go back to the board and come up with a revision.

Want to bet?

FRED POPPE

Acknowledgments

To all the people below, my heartfelt thanks for helping me put together this tome. Also a very special thanks to Vye Messner for her superb help in keeping me from going ahead with that prefrontal lobotomy which I so richly deserved:

Wally Armbruster
Al Bugbee
Wade Cloyd
Ron Coleman
Bill Cowan
Art D'Arasian
John DeWolf
Bob Donath
Keith Evans
Bob Foley
Jim Gillam
Fred Harvey
Ed Hatcher
Ed Hughes
Ray Johnson
Richard Kerr
Bob Lukovics
Tim McGraw
Fred Messner
Fergus O'Daly
Steve Rand
Norman Richardson
Lee Rosen
Scotty Sawyer
Joe Serkowich
Bob Singer
The Copy Chasers
Larry Wattman
Walter Weir
Bob Welborn
All my clients

Contents

Foreword: So-Called Industrial Advertising/vii
Preface/ix
Advertising Council/3
Alcoa/5
Amchem/7
American Mutual/9
American Optical/11
American Telephone & Telegraph/13
Andersen Corporation/15
Armco/17
Associated Spring/19
Bethlehem Steel/21
Brenco/23
Cadillac/25
Calgon/27
Carborundum/28
Celanese/31
Chemical Engineering/33
Clark Equipment/35
Container Corporation of America/37
Diamond Shamrock/38
Downs Crane & Hoist, Inc./41
Du Pont/42
Durkee/45
Eastman Kodak/47
Elastic Stop Nut/49
Fafnir/50
Federal Express/53
Fireman's Fund/55
Forbes/57
Foxboro/59
GAF/60
Gates/62
General Electric/64
B.F. Goodrich/67
Goodyear/69
Hercules/71
Hewlett-Packard/72
Hoffman-LaRoche/74
Holo-Krome/76
Homequity/Homerica/79
Honeywell/81
Hughes Aircraft/83
IBM/84
INA/86
Inland Steel/88
International Harvester/90

International Paper/92
ITT/95
J. & L. Steel/96
Johnson Wax/99
Jones & Lamson/101
Le Tourneau-Westinghouse/103
Lima Hamilton/105
Loctite/107
Lonestar/109
Lukens Steel/110
Marsteller/113
McGraw-Hill/115
Mobil/117
Motorola/118
National Starch/120
New York and Penn/123
Norton/125
Olin/126
Owens-Corning/129
Peterbilt/130
Pitney-Bowes/133
Polaroid/134
PPG Industries/136
Raybestos/139
Raytheon/140
Republic Aviation/143
Rockwell/145
Rockwell Spring/147
Rome Wire Company/149
Ryerson/151
Sears/152
Shell Oil Company/155
SKF Industries/157
St. Regis Paper Company/158
Sonoco/160
Spectra-Strip/162
Sperry/164
Stewart & Stevenson/167
Sweco/169
Swingline/171
3M/172
Timken/174
United Shoe Machinery—Nylok/177
United Technologies/179
U.S. Industrial Chemicals/181
U.S. Steel/182
Veeder-Root/185

Volkswagen/187
Wang Laboratories/189
Warner & Swasey/191
Westinghouse/192
Wheeling Steel/194
Wickwire Spencer Steel/197
Xerox/198
Zilog/200

THE 100 GREATEST CORPORATE and INDUSTRIAL ADS

Pollution: it's a crying shame

**People start pollution.
People can stop it.**

Keep America Beautiful

Advertising contributed for the public good

Advertising Council

The Advertising Council is a nonprofit organization through which American business and the advertising and communications industries contribute their skills and resources to promote voluntary citizen action in solving national problems. In the early 1950s I was with G.M. Basford Company, a volunteer agency, and got involved writing business paper ads to help sell war bonds (by then they were calling them U.S. Treasury Defense Bonds). Our ads were designed to convince business management to set up payroll savings plans for employees to make it easier for them to save by investing in U.S. Bonds. We'd prepare the ads in repro form and offer them to some 2,500 business publications. The most public spirited of the trade press would then run the ads free of charge. Over the years, millions and millions of dollars worth of free space was donated to this program. And naturally, billions of dollars worth of savings bonds were sold through payroll group plans.

One of the greatest of The Advertising Council's campaigns is this "Keep America Beautiful" series. Marsteller Inc. is behind this very special ad which features a sad American Indian (Iron Eyes Cody) with a tear streaking down his face. Marsteller also furnished videotaped TV spots for the country's television stations which donated free air time. The crying Indian TV commercial was aired more often than any other commercial in the history of television. It featured Mr. Cody paddling his birchbark canoe through a garbage-infested river. As he paddled by empty beer cans, the tear descended from a doleful eye down along his fine aquiline nose. The theme, "People Start Pollution. People Can Stop It," was in both the television and print copy. The grabber here, of course, is that wonderful photograph of this very authentic American Indian. Whoever cast the model and was behind the camera deserves a special kudo. Marsteller's Hugh McGraw wrote the copy; Joe Goldberg was the art director.

HOW A MODERN BUS LOOKS
without Cosmetics

UNDER THE RAINBOW COLORS of the modern bus is the most significant fact in transportation: The strong alloys of Aluminum are the Safe Way to Lightness.

Buses *have* to be light, these days. Dead load is expensive to haul around. Operators have to have buses spacious enough to carry many passengers in comfort, yet light enough to be agile in starting and economical to run.

From the very beginning of the trend to lightness, bus builders and operators have recognized that the need for both lightness *and* safety is a clear call for Aluminum.

Nature made Aluminum light, and research has made it strong. With this unbeatable combination, builders can throw off thousands of pounds of weight and still have a structure that is inherently stiff, dependably rigid, strong, and able to absorb the impacts of service.

Every important bus builder in America uses this safe way to lightness. They build throughout of strong Aluminum Alloys, in the form of rolled and extruded shapes, sheet and rod, castings and forgings. They build economically, because Aluminum is easy to fabricate. Ordinary shop techniques are quickly adapted to handle this versatile metal.

Operators get, with Aluminum, not only weight saving with safety, but a bus that is easy and economical to maintain, and that has a high salvage value when the equipment becomes obsolete.

A LIFTABLE IDEA: ALUMINUM IS THE SAFE WAY TO LIGHTNESS

The whole transportation industry has adopted this simple idea. The airplane is Aluminum; buses, you have just read about. Railroads use this safe way to lightness for passenger cars, tank cars, and other equipment. Streetcars, too. Truck bodies and truck tanks are lightened with Aluminum. Bicycles and outboard motors. Gas engine and Diesel engine pistons and other parts.

Remember: Aluminum is the safe way to Lightness. It's a liftable idea, which we can help you put into practice, economically. Aluminum Company of America, 2102 Gulf Building, Pittsburgh, Pennsylvania.

THE *Safe* WAY TO LIGHTNESS

ALCOA · ALUMINUM

This advertisement No. ES-8318 appears in Fortune, Nation's Business, Manufacturers Record, October, Business Week, October 1st and U. S. News, September 19th, 1938.

Alcoa

You have to give Alcoa credit for its consistency. The copy platform for both these great ads sells the lightweight strength and low-maintenance benefits of aluminum to the transportation industry. "How a Modern Bus Looks Without Cosmetics" ran in 1938 in *U.S. News & World Report,* and "This glistening bird could use less fuel" ran in 1981 in guess where? You've got it. *U.S. News & World Report!* And the advertising agency that prepared the 1938 ad was Fuller & Smith & Ross. The agency that prepared the 1981 ad 43 years later was Creamer Inc., the agency that merged with Fuller & Smith & Ross a few years back. Dale Worcester, Alcoa's manager of marketing communications, who is also in charge of corporate advertising, credits this consistent advertising for helping transportation to become the third-largest market for aluminum today. Both ads use four colors beneficially to dramatize and add stopping power to the overall first impression the reader gets as he's thumbing pages. That tiny rabbit running in front of the bus in the 1938 illustration, for instance, is a fine touch of whimsey that helps humanize the ad and make it come alive. The two jet streams of water hitting the 727's bright skin is highlighted in the 1981 photograph in just the right way to snare more scanners and turn them into readers. No wonder Alcoa has been the spokescompany for the entire aluminum industry.

If it weren't for Amchem, he'd have to settle for a '72.

Because his classic automobile is made of steel.

And steel rusts.

And Amchem came up with the first pre-paint conditioner that made steel auto bodies feasible—back in 1914 when rust was coming through as many as 15 coats of paint.

Later, Amchem developed the first cold phosphating process for steel. Then the first conversion coating that got the aluminum can and painted aluminum industries started.

Today, we're *still* coming up with firsts. In fact, research and development takes up more space at Amchem than any other division except manufacturing. And what we're doing in that space, for example, is working on new chrome-free and phosphate-free processes. New techniques for treating waste water. Whole new systems for better use of existing water.

And at Amchem we know that, today, *service* is more important than ever. Experienced, technical service that includes in-plant engineering and design assistance. Startup guidance, and maintenance training for your people. Quality control services, and short-notice troubleshooting.

Being first yesterday isn't enough. Being first with you, today, is what we're after. For everything in pre-paint chemicals for metal, call your Amchem representative, or write direct. Amchem Products, Inc., Ambler, Pa. 19002.

Amchem

Amchem Products, a division of Henkel Inc., has been running a series of these four-color single-page bleed advertisements for over nine years. They're aimed at industry in general and feature Amchem's leadership in producing prepaint chemicals for metal protection. This ad is the very first of a number of "old car" ads prepared by the Lewis & Gilman ad agency in Philadelphia. Obviously, it ran in 1972, as the stopper headline succinctly points out. Since its inception, the campaign has been a consistent winner in readership studies. Many Best in Book and Best in Product Category ratings attest to the effectiveness of this unique series. The editors of a number of the publications in which the ads ran liked them so much they gave Amchem thousands of dollars worth of cover positions gratis. Knowing how parsimonious publishers are wont to be, I'd say that in itself earns the Amchem ad distinction as another one of the "greatest."

American Mutual

Despite a number of surveys proving that cartoon ads do not pull as well as conventional ads, there are some exceptions. In the 1950s American Mutual Liability Insurance Company ran a series of lovely cartoon ads featuring Mr. Friendly as the "eyepatch." They ran in *Business Week, Wall Street Journal, Newsweek* and *Factory*. This one, entitled "The Case of the Suspicious Businessman!," was honored in 1954 by Business Week as being one of the highest-scoring ads of the year. Mr. Friendly was a character drawn by that creative giant Jack Tinker when he was at McCann-Erickson. In every ad in the series, Mr. Friendly was dressed in a rather funereal way. He always wore a black bowler hat and more often than not carried a tightly rolled umbrella. Tinker was a master at showing all kinds of frenetic activity in his drawings which of course engendered great readership.

This "greatest" starts with a poem about how businessmen should be more suspicious about their employees as well as outsiders. Mr. Friendly then very eloquently espouses American Mutual's Comprehensive Crime Policy. A separate sign-off piece of more serious copy implores the reader to send for a free booklet entitled "Modernizing Your Crime Loss Insurance Protection," and it's all bottomed off by a neat-looking AM logo adorned with a line drawing of Mr. Friendly, bowler, umbrella, and attache case. Credit for the copy goes to Don Calhoun, also of McCann.

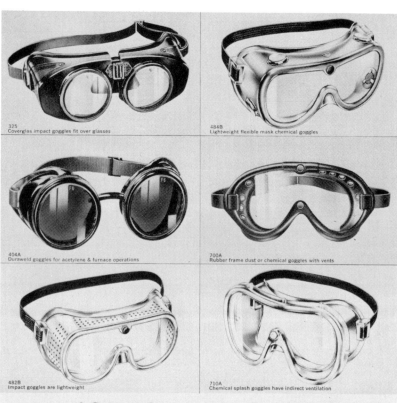

AO goggles protect the eyes that do

And these are only the beginning. Only a few from more than 200 different types in AO's complete line.
(Where the job demands it, you also have a choice of over 1500 pairs of safety glasses and 220 face shields. So, if it's something for the eyes, you'll find a complete protection package at American Optical.)
All these safety goggles are designed to give you the best in protection plus maximum vision. Add comfort, and you have everything a pair of goggles should be. Plus the fact that they're made to stand up under tough service. Parts are easily replaced.
And there's a selection of lenses that is unequalled for any job. For impact protection, you have Super Armorplate® (clear or Calobar®), Plastolite®, and other plastic lenses (in clear or green) which also protect against chemical splash and spray. And the welder can feel secure in shades 4, 5, or 6 of Filterweld® lenses.

® ™ Registered by American Optical Corporation

American Optical

Let's face it: goggles aren't sexy. You can sell millions of designer sunglasses with a print ad showing a pneumatic Raquel Welch hiding benignly behind her Foster Grants, but how in hell do you get safety engineers to specify dull and, very often, ugly safety goggles? Just a dozen years ago the people at American Optical came up with the answer. Wade Cloyd, the advertising maven at AO, got together with Al Scherm, the Fuller & Smith & Ross account executive, and his art director, a taciturn Scotsman named Jim Robertson. The three of them came up with this beautiful four-color gutter-bleed spread with a pictorial portion divided into 12 horizontal rectangles. Just off-center, but right at the ad's focal point, is a pair of wide-open, unprotected eyes that literally forces the reader's eyes right to the

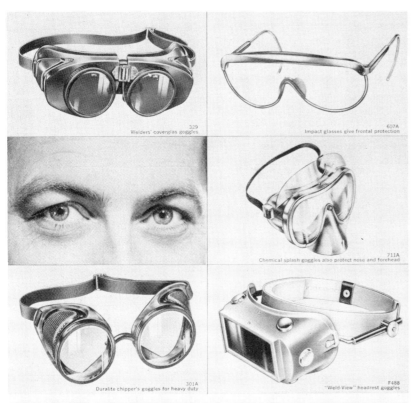

the job. Any job.

Both the flexible mask and solid cup types fit comfortably over all personal and safety Rx glasses. The flexible masks of AO chemical goggles, either vinylite or neoprene rubber, fit the contours of the face. Welding goggles have indirect ventilation which lets in cool air but not harmful rays or particles.

Whatever the eye hazard, your AO Safety Representative can recommend countermeasures. Plus cleaning and antifog solutions and germicidal detergents for sanitation.

When you need goggles or safety glasses, respirators or head and body protection, look to AO SUREGUARD products for your surest protection. Make AO your SAFETY SERVICE HEADQUARTERS.

AMERICAN OPTICAL CORPORATION
Safety Products Division • Southbridge, Massachusetts 01550

page! In the other 11 spaces are brilliantly depicted pairs of goggles that AO selected as best representing its broad line of more than 200 types of goggles. A brief caption including a model number accompanies each set of goggles. (We know that's smart because photos with captions always outpull photos without captions.) The results of the ad are now legendary. For seven or eight years, every time the ad ran, it brought in a slew of orders for safety goggles. Each order included the particular model number shown in the ad. Also, every time it ran, it racked up top readership study awards in such trade publications as *Chemical Week, Iron Age, National Safety News, American Machinist* and many others.

It's nice to assume that over the years this ad has helped avert at least some of the estimated 140,000 work-related eye injuries that occur in the United States every year.

"The Voice with a Smile"

"Hail ye small, sweet courtesies of life, for smooth do ye make the road of it."

Often we hear comments on the courtesy of telephone people and we are mighty glad to have them.

For our part, we would like to say a word about the courtesy of those who use the telephone.

Your co-operation is always a big help in maintaining good telephone service and we want you to know how much we appreciate it.

BELL TELEPHONE SYSTEM

American Telephone & Telegraph

The Bell Telephone System advertising account has been at N.W. Ayer since 1908. This has to be the longest agency-client relationship in the history of advertising. The Bell System campaign also has to be one of the most consistently great corporate and institutional campaigns ever run. The utter simplicity of "The Voice with a Smile" is really what makes it one of the best. For years the telephone company has been paranoid about its image as a monopoly. In order to allay the attitude that "big is bad" and "let the public be damned," its advertising has had to instill confidence and credibility to all of its many publics.

Now, who would dare not trust that pretty lady with the smiling voice? I'd damn motherhood, Nancy Reagan or Robert Redford before I'd blaspheme anyone so saintly as this Ma Bell's little helper. Brad Lynch, Ayer's vice president corporate communications, couldn't come up with art and writing credits, but in all probability this one should be dedicated to George Cecil, Ayer's late, great copy chief in Philadelphia, who shepherded over the phone company's advertising for more than 50 years.

One way to insulate.

The best way to insulate with a view.

Weathertight Andersen® Windows have always let you make the most of a beautiful view.

(And that's a lot better than no windows at all!)

What you may not know is that Andersen Windows are designed to cut your fuel bill.

Andersen Windows are made of wood, one of nature's best insulators.

And they're built two times more weathertight than industry standards.

With optional double-pane insulating glass, Andersen Windows also cut heat loss through the glass. Eliminate storm window bother, too.

So save fuel *and* enjoy your view. Insist on Andersen Primed Wood or Perma-Shield® vinyl-clad Windows.

For further details, see your lumber dealer. He's in the Yellow Pages under "Windows, Wood."

Please send me your free booklet, "How to Get Good Windows."
☐ I plan to build.   ☐ I plan to remodel.

Name _____

Address _____

City _____ State _____ Zip ____

The beautiful way to save fuel.

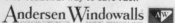
Andersen Windowalls
ANDERSEN CORPORATION BAYPORT, MINNESOTA 55003

Andersen Corporation

Industrial Marketing selected this full-color Andersen Windowall's effort as one of the best industrial ads that ran in 1975. The message is treated so simply you virtually don't have to read the copy. The first house is insulated beautifully because it has no windows. The same house with Andersen windows cut in is also insulated, but this time the house has a view. How can a selling proposition like this miss? Not at all! Bob Warren, the copywriter, and art director Dave Bradley of Campbell-Mithun deserve all the credit for putting this one together. Text is set in three narrow columns with margins justified flush-left and flush-right for easy reading. The headline is set in uppercase and lowercase and separated in order to serve as captions for the two pictures. The caption-headline, incidentally, is a device that should be used more in business-to-business advertising because it is known to increase readership. The tag line, "The beautiful way to save fuel," is one of the few slogans around today that not only makes sense, but also adds to the appeal and pulling power of the ad.

"Dutch" Fulton
TURNS ON THE HEAT!

HEATING electrical sheet steel is a pretty ticklish job, as "Dutch" Fulton could tell you. Dutch has been at it for thirty years in ARMCO's up-to-date mill. When your order comes along, he knows from long experience just what to do with it.

Let's follow "Dutch" for a minute... First, he is responsible to the roller for heating the sheets properly before rolling. But the roller doesn't need to watch our friend "Dutch" closely, because "Dutch" knows his work. Now he closes a damper; now he opens it. A deft touch of the fuel regulator turns on more heat—or less—whatever the sheets require. In and out of the glowing furnace go "Dutch's" tongs—and every time without fail the roller gets steel heated properly for efficient rolling.

It's an interesting job and "Dutch" likes it for the sheer joy of working with other experienced men and producing fine electrical sheet steel. He has known the rest of his crew for years—so long, in fact, that they can literally "smell" trouble before it arrives and change the condition to meet it. You will readily understand what this means to punching and stacking in your shop, to say nothing of magnetic properties that insure top-notch efficiency of your electrical machinery.

ARMCO can give you everything that contributes to highest-quality electrical sheets —experience, research facilities, modern mill equipment, and painstaking care. We invite your problems and requirements. The American Rolling Mill Company, Executive Offices, Middletown, Ohio.

ARMCO  ELECTRICAL SHEETS

Armco

Industrial advertisers had the same problems back in 1936 as we do today. To quote that year's *Industrial Marketing:* "The magazines are getting thicker (we begin to loathe 'special issues')—it grows harder and harder to distinguish the very good from the good—the old eyes blink with advertising astigmatism." The ad of the month picked by Copy Chasers 46 years ago quite deservedly is this Armco (then the American Rolling Mill Company) ad written in-house by William McFee. "Dutch Fulton turns on the Heat" is a tribute to McFee's copywriting acumen. How many "real people" ads have been turned down by reticent advertisers because management feared the featured person might leave his job, or sue the company, or whatever? The fact is that real people get readership and believability and that's the name of the game. Dutch Fulton wasn't hired by a pretty boy model agency. He's a tough-looking steelmaker character who, as *Industrial Marketing* pointed out, "wasn't asked to straighten out his tie or watch the birdie." He's busier than busy with his job. And what's his job! The copy tells you that:

> Heating electrical sheet steel is a pretty ticklish job, as Dutch Fulton could tell you. Dutch has been at it for thirty years in Armco's up-to-date mill. When your order comes along, he knows from long experience just what to do with it.
> Let's follow "Dutch" for a minute...First, he is responsible to the roller for heating the sheets properly before rolling. But the roller doesn't need to watch our friend Dutch closely, because Dutch knows his work. Now he closes a damper; now he opens it. A deft touch of the fuel regulator turns on more heat—or less—whatever the sheets require. In and out of the glowing furnace go Dutch's tongs—and every time without fail the roller gets steel heated properly for efficient rolling.

Those are the first two paragraphs of copy that is vivid and alive, human and interesting. It carries along with the same punch into a "duality product" story—effective and rememberable. It's a swell job.

Inside Stuff

Service by Design

ABOUT SPRINGS, CUSTOM METAL PARTS AND ASSEMBLIES

Begin at the End in Extension Spring Design

AG 41

When designing extension springs the stresses in hooks and loops must be carefully considered along with all other specifications. The reason is that these stresses, which are *both* bending and torsional when the spring is extended, frequently exceed the stresses in the body of the spring. This becomes even more significant when the hooks or loops may be subject to both systems of stresses in combination.

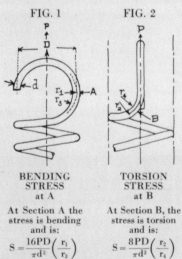

BENDING STRESS at A

At Section A the stress is bending and is:

$$S = \frac{16PD}{\pi d^3}\left(\frac{r_1}{r_3}\right)$$

TORSION STRESS at B

At Section B, the stress is torsion and is:

$$S = \frac{8PD}{\pi d^3}\left(\frac{r_2}{r_4}\right)$$

Critical stresses can occur at sections A (Fig. 1) and B (Fig. 2). Recommended practice is to make r_4 greater than $2d$. This can be alleviated by winding the coils with decreasing diameter so that hook stresses are reduced (Fig. 3).

Assembled hooks are another fix. A flat stamping (Fig. 4) or a threaded plug (Fig. 5) reduces stress concentra-

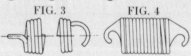

FIG. 3   FIG. 4

tions by distributing the load over several coils of the body. A swivel-hooked assembly (Fig. 6) also can be used to reduce stress because the flexible joint eliminates or reduces the bending stresses of the hook, leaving only the torsional stress to consider.

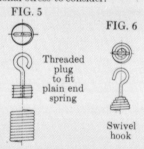

FIG. 5   FIG. 6

Threaded plug to fit plain end spring

Swivel hook

A convenient way to adjust spring rate, or the frequency of vibration of the spring system, is to vary the number of active coils. With the flat stamping (Fig. 4) or threaded plug (Fig. 5) this can be done by simply screwing or unscrewing the stamping or plug.

The maximum extended length without set of an extension spring must be calculated and specified because an extension spring does not have a definite deflection stop, as does a compression spring.

If you're embarking on a project requiring springs, or any metal component, let us show you how A.S.C. technical and production short cuts can help. *For greatest convenience, ask for the location of the A.S.C. facility nearest you.*

6906

Associated Spring Corporation

Corporate Office
Bristol, Conn. 06010

Plants throughout the U.S.A. and in Canada, Mexico, Argentina, England, The Netherlands

Associated Spring

Who says industrial advertising isn't a science? When the job is to engage engineers, sometimes you've got to sell a lot more than just the sizzle of the steak. You've also got to appeal to their engineering sense, and to do that, you have to talk to them in the patois of their profession. And that's exactly what Kent Putnam, the former ad manager of Associated Spring, did when he came up with this now famous 2/3-page black and white ad campaign titled "Inside Stuff." For years, Associated Spring had been publishing a design manual to help inform engineers on how to best solve their spring problems. The manual was so highly regarded it was known as the bible of the industry. That gave Kent the bright idea of excerpting important sections of it to use in his advertising campaign. His agency, Davis Advertising of Worcester, Massachusetts, designed an attractive masthead to give each ad continuity and the overall look of an informative newsletter, which research told them would garner readers by the score. In almost pure engineering jargon, the copy invites readership by educating the reader about things like "torsional and bending spring stresses" and then showing these stresses in handsome engineering diagrams. This kind of advertising acts like catnip to the engineering purist. In the magazines this campaign appears in, the median readership Ad Gage scores for *full-page* black and white ads runs 8%. The average ratings Associated has been earning since this 2/3-page campaign was conceived amounts to 40% to 50%. That means these great ads are pulling more than 600% above the average every time they appear. And that's why Associated Spring is still running them 15 years later.

TOE-HOLD ON A BRIDGE

A striking example of Bethlehem's ability to meet unusual wire-rope requirements

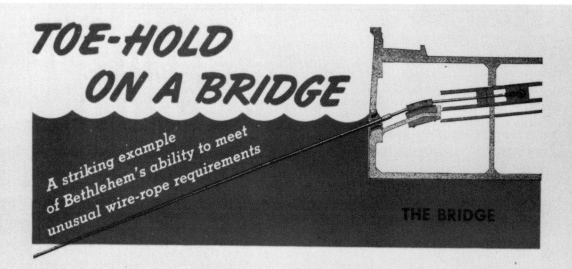

THE BRIDGE

Building the Mercer Island Bridge, longest pontoon bridge in the world, across deep, fresh-water Lake Washington created a number of unusual engineering problems. Outstanding was the problem of anchoring the bridge. Cables had to be used to preserve the flexibility of the structure. These cables had to be of high strength, yet had to be protected against corrosion.

Bethlehem detailed and fabricated the anchor cables for the Mercer Island Bridge. There are 64 of them, ranging up to 614 feet in length. They are 2¾ in. in diameter, 6 x 37 construction, with an independent wire-rope center. They meet a 620,000-lb. strength specification, and on test easily passed the specified strength.

The patented bethanizing process was used to zinc coat the wires of these cables. This electrolytic process, used exclusively by Bethlehem, builds up a tightly bonded, uniform zinc coating without affecting the strength or toughness of the cold-drawn wires in any way. It would have been impossible to produce a rope of this strength and toughness by any ordinary galvanizing process; equally impossible to produce a uniform hot-galvanized rope that met the coating-weight specification: 1.6 ounces per square foot of wire surface compared to 0.72 for conventional high-grade galvanized rope.

These cables represent a very unusual and very specialized job. They are a convincing demonstration of Bethlehem's ability to take such jobs in stride. With modern equipment, backed up by one of the finest metallurgical and mill organizations in the world, Bethlehem is prepared to help you on any problem involving wire rope.

AN ANCHOR

BETHLEHEM STEEL COMPANY

Bethlehem Steel

John DeWolf, the venerable copy sage who invented *visual magnetism* (a way to forecast readership scores), wrote this gem of an ad for Bethlehem Steel. The ad appeared in *Engineering News Record* over 40 years ago, and with some modern, up-to-date typography, it could be run again today and be just as effective.

The provocative headline "Toe-Hold On a Bridge" and the simplistic, almost diagrammatic artwork tell the whole tale in an instant. The copy is a factual story, brilliantly told, which outlines the problems involved in anchoring the Mercer Island Bridge. The fact that it happened to be the world's longest pontoon bridge made the story even more adventurous, and DeWolf's copy played it to the hilt. He very tersely describes the problems and then gets down to the nitty-gritty by explaining how Bethlehem licked these problems through its specialized and exclusive bethanizing process. The trick here was to do all of this in a convincing tone. And that's exactly what the ad does. The engineering/contractor reader is "talked to" in the argot of his profession. The copy also emulates the editorial style of *Engineering News Record* and in so doing adds a great deal of believability to the message. This is another fine example of a non-ad approach that works mainly because it's written in a noncommercial way.

An open letter to railroad management and to the railroad's energy-conscious customers.

"...if you could save 30 gallons of fuel for every hour of operation of a 100-car freight train... merely by specifying the right seal... would you do it?"

Gentlemen:

A small, but vital part of your freight cars is costing you more in fuel than you are probably aware: the roller bearing seal.

It takes a significant amount of energy—fuel—to operate the roller bearing, because the lubrication seal creates friction or drag when it does its job properly.

Unfortunately, some seals consume more energy than others.

It has been the goal of Brenco roller bearing seal research to lessen this energy consumption to a minimum and to improve seal life. We feel that our latest developments have achieved results that you should be aware of. They will enable you to make significant improvements in fuel consumption by freight trains. Here are some of the highlights of a recent study of the new Brenco seal, with hydrodynamic characteristics.

For example...

If you could save 30 gallons of fuel for every hour of operation of a 100-car freight train... merely by specifying the right roller bearing seal... would you do it?

We are inclined to think you would. Here are the facts.

Consider a 100-car freight train equipped with the competitive roller bearing seal in use on the majority of freight cars today. The locomotives will burn approximately 100 gallons of fuel every hour, at 50 mph, just to overcome the friction of these seals. This is in addition to the fuel needed to pull the rest of the load.

If the same train were equipped with the new Brenco seal, which requires less energy than all competitive seals, the train would consume 30 gallons of fuel an hour less!

If the performance of the Brenco seal is compared to the average of all the competitive makes available, these results are obtained:

The 6½ x 12 Brenco seal consumes 18 percent less fuel than the average of all the others.

At 34¢ a gallon for fuel, this equals a savings of about $1.00 for every 1000 car-miles.

Better operation too. A Brenco seal has the advantage of both low wear and high sealing effectiveness. And most importantly, the Brenco seal retains these qualities and has a longer effective service life than all other competitive makes.

Fuel Consumption Comparison
Fuel Consumption-
Gallons/1000 car-miles, new seals.

Competitive Seal "A"*: 20.2
Average of five Competitive Seals: 16.8
Brenco Seal: 13.9

*This seal used on the majority of cars rolling today.

...Example: after 65,000 miles of operation, the 6½ x 12 Brenco seal is 12 percent lower in energy consumption than its best competitor. And at the same time the Brenco seal is also *61 percent more effective* in its ability to seal the bearing—as measured by residual shaft interference fit.

As one railroad mechanical officer remarked, "If that's so, it's like having your cake and eating it too!"

Brenco seal technology is at the heart of this development. Brenco is the only freight car roller bearing manufacturer to make its own seals. The technology and the equipment that we have developed enables us to maintain rigid quality control over our seal every step in its manufacture.

We do this because we know the reliability of the bearing depends upon the seal.

And because our name is on our product.

Brenco has manufactured more than three million seals. Their record of performance has been outstanding. We are proud of that record.

We are also proud of our latest roller bearing seal development, which has an advanced lip geometry and molding techniques. The savings in fuel that this seal makes possible will be available to the railroad industry, not only in our own bearings, but through the AAR approved interchange program, in all bearings, new or reconditioned... wherever Brenco seals are specified.

W. Stuart Johnson
President, Brenco Incorporated

For more information on Brenco seal research, write for our technical report "New Fuel Economies Through Roller Bearing Seals."
Brenco Incorporated, P.O. Box 389, Petersburg, Virginia 23803.

**Brenco tapered roller bearings
Brenco seals**

BRENCO

THIS ADVERTISEMENT APPEARED IN THE WALL STREET JOURNAL, APRIL 5, 1976.

Brenco

Why are long-copy, all-type ads anathema to most copywriters and agency account executives? Quite frankly because they are difficult to write, and, even more significantly, they're almost impossible to sell to the client, especially client management. Most every client president I've ever worked for is an advertising expert. They seem to enjoy dabbling in writing, designing or even creating ideas for ads and ad campaigns. Ninety-nine percent of them hate long copy because they feel nobody (as busy as they are) has time to read a long advertisement. They're wrong. This Brenco open letter ad, written by Spectrum Marketing's Oliver Darling, is another sterling example of how long copy can work. Written in the tradition of the best-selling Sherwin Cody mail order ad, "Do you make these mistakes in English?," it proved again that good business-to-business advertising can sell products. Just three months after its appearance in a bunch of railroad magazines and the *Wall Street Journal,* sales of Brenco's tapered roller bearing seals tripled.

If you think railway purchases and stores people think a bearing is a bearing is a bearing, not so! After you read Darling's informative, easy-to-understand treatise, you find out that at 34¢ a gallon for diesel fuel (back in 1976), Brenco roller bearing seals achieved savings of about $1.00 for every 1,000 car-miles. That's real fuel economy, and this fine ad spells it out beautifully without having to show a big illustration of a train to attract readers.

THE PENALTY OF *Leadership*

In every field of human endeavor, he that is first must perpetually live in the white light of publicity. Whether the leadership be vested in a man or in a manufactured product, emulation and envy are ever at work. In art, in literature, in music, in industry, the reward and the punishment are always the same. The reward is widespread recognition; the punishment, fierce denial and detraction. When a man's work becomes a standard for the whole world, it also becomes a target for the shafts of the envious few. If his work be merely mediocre, he will be left severely alone—if he achieve a masterpiece, it will set a million tongues a-wagging. Jealousy does not protrude its forked tongue at the artist who produces a commonplace painting. Whatsoever you write, or paint, or play, or sing, or build, no one will strive to surpass or to slander you, unless your work be stamped with the seal of genius. Long, long after a great work or a good work has been done, those who are disappointed or envious continue to cry out that it cannot be done. Spiteful little voices in the domain of art were raised against our own Whistler as a mountebank, long after the big world had acclaimed him its greatest artistic genius. Multitudes flocked to Bayreuth to worship at the musical shrine of Wagner, while the little group of those whom he had dethroned and displaced argued angrily that he was no musician at all. The little world continued to protest that Fulton could never build a steamboat, while the big world flocked to the river banks to see his boat steam by. The leader is assailed because he is a leader, and the effort to equal him is merely added proof of that leadership. Failing to equal or to excel, the follower seeks to depreciate and to destroy—but only confirms once more the superiority of that which he strives to supplant.

There is nothing new in this. It is as old as the world and as old as the human passions—envy, fear, greed, ambition, and the desire to surpass. And it all avails nothing. If the leader truly leads, he remains—the leader. Master-poet, master-painter, master-workman, each in his turn is assailed, and each holds his laurels through the ages. That which is good or great makes itself known, no matter how loud the clamor of denial. That which deserves to live—lives.

This text appeared as an advertisement in The Saturday Evening Post, January 2nd, in the year 1915. Copyright, Cadillac Motor Car Company

Cadillac

This corporate advertisement is the granddaddy of them all. It appeared in the January 2 issue of the *Saturday Evening Post* in 1915. It's all copy and it's long copy. Some 400-plus words. But what copy! The Penalty of Leadership is not "widespread recognition" and the "white light of publicity," but "fierce denial and detraction." When leadership is vested on a man or product "emulation and envy are ever at work." This is a beautiful use of action words to get a point across. Read further. "When a man's work becomes a standard for the whole world, it also becomes a target for the shafts of the envious few." Just one line of copy in a corporate ad, but it could have been lifted from or added to Lincoln's *Gettysburg Address*. It reads as if its author were a professor of humanities or at least held a doctorate in English literature instead of being an advertising agency copywriter. The ad was written by Theodore MacManus of MacManus Advertising and is considered *the* classic corporate advertisement of all time. Even though it has passed its sixty-sixth birthday, this ad still hangs in the office of many corporate executives. Its all-copy format is set in a hard-to-read script type face, but it would have been a sacrilege to have set it in any other face. The entire ad takes almost four minutes to read thoroughly (you could view eight TV commercials in that span of time). But you have to believe that this is another one of the best arguments against those who scream about the use of long copy in corporate advertising.

Would you bathe a baby in secondhand water?

You would.

And do.

The clean water that bathes a baby today could have cooled a motor, quenched a fire, watered a rose garden, washed dishes, made paper, tempered steel, iced a drink or even bathed another baby only a day or two ago. More and more water is used water.

The challenging problem of today is keeping used water clean enough to use again. In more than 40 years of helping industry meet that challenge, Calgon Corporation has discovered there is no substitute for sound, experienced engineering when it comes to doing that job effectively and at realistic cost.

Learn how you can help industry and government in your area work towards sound, clean-water objectives—and perhaps how Calgon can help you. Write for "The Challenging Problems of Water," Calgon Corporation, Dept. G, Calgon Center, Pittsburgh, Pa. 15230.

Helping America answer the challenging problems of water

1-522   HC-62872

Calgon

Business and professional ad writers aren't stupid. We have to know just as much as our consumer friends about what appeals most to our readers. Very often we have to flavor our technical lingo with saccharine in order to make it palatable to our audience. We're smart enough to know that the same appeal can be used to sell either an expensive elevator system or a pack of cigarettes. One of the first things a cub copywriter learns is the magical pulling power of babies, horses and dogs. When in doubt, put an animal in an ad and it will get as much attention (and therefore pulling power) as the words **NOW, NEW, FREE** and **ANNOUNCING** do. In 1965, this fine black and white Calgon ad was the first in a series designed to call attention to Calgon's capabilities in pollution control and water preservation. Calgon's agency, Ketchum, MacLeod & Grove, was smart enough to launch it using a happy cherub above the admonishing headline, "Would you bathe a baby in secondhand water?" The ads ran a full page in the Sunday *New York Times, Harper's, Atlantic Monthly* and other thought-leader publications. Calgon has received thousands of inquiries for its booklet "The Challenging Problems of Water," even though the offer for it was well-hidden in the body text. Obviously, people were reading the entire message. By sheer coincidence, the first ads broke during the summer in which New York City experienced a dreadful water shortage for the first time in years. This bit of luck added a tremendous amount of interest and gave vigorous impact to getting the campaign rolling. Now that's some timing!

Carborundum

In 1965 the Carborundum Company was nowhere insofar as recognition was concerned. Most of the advertising that was being done was strictly product-oriented, and what little there was represented a holding-action strategy. Then out of the blue, Frank Massard, Carborundum's manager of advertising, working with Rumrill Hoyt, developed a corporate campaign that eventually begat an advertising renaissance that gave the company an entirely new look. Research informed Gene Novak, Rumrill Hoyt's creative director, that Carborundum's publics thought of the company as an unsophisticated supplier of grinding equipment. The name itself is a tongue twister and does nothing but convey a sandpaper image. Rumrill's full-color campaign changed all that forthwith. Using superb photography for top impact on the left-hand page, and stark black

We've invented a match you can strike 250,000 times.

Until now,
the billion dollar
gas appliance and
equipment industry
has been nagged by a problem.

How to light the burner.

Matches are primitive.
Pilots blow out.
Spark gaps plug.
Coils burn out.

Today,
there's an answer crafted by Carborundum's
capability in materials and material systems.

Our invention is the first
transformerless resistive igniter.
It is made of silicon carbide and other
Carborundum ceramics. It will light a burner
reliably 250,000 times. The quantity price
is less than $3.00.

Carborundum's igniter is already in
gas dryers that you can buy today.
We think it soon will be in gas furnaces,
range ovens, boilers, infrared heaters.
Just about any place a gas flame
has to be lit reliably.

We're no stranger to appliances.
Carborundum varistors and thermistors
degauss 60% of today's color TV sets.
Our Fiberfrax® ceramic fiber cement
and insulation are in the
Corning Counter That Cooks.

Our materials capability will go
wherever we can help do something
better than it's been done before.

Call Mr. Robinson.
If a gas igniter, or anything else
using our materials interests you for
appliances, Mr. K. T. Robinson,
Marketing Manager, Electronics,
will appreciate your call
at 716-278-2546.
The Carborundum Company,
Niagara Falls, N.Y. 14302

CARBORUNDUM

---

and white text and white space on the right hand page, Novak's copy and art team came up with a spread series that ran in *Business Week* for years. The basic objective was to increase awareness among customers, prospects, security analysts and businessmen in general of Carborundum's broad range of products and capabilities.

My favorite, "We've invented a match you can strike 250,000 times," was a stunning example of how well these ads fared insofar as readership was concerned. In one year no ad ranked less than fifth, and this match ad was the highest scorer for all ads studied in *Business Week* for the year. Two others in the campaign ranked second.

Benchmark research proved the campaign was a total success: awareness increased, calls from security analysts poured in and mailings to top executives produced a plethora of responses.

growth rings

A sound idea spreads like rings on a pond. Each geometrically larger, reaching farther than the one before.

We can trace our own rapid growth to Celanese men and women of free-wheeling imagination—who hone their ideas on the exacting and rigorous disciplines of modern chemistry.

They've helped us add new products . . . improve existing ones . . . develop unique processes . . . build new plants to serve rapidly expanding markets at home and abroad.

So we search diligently for people with the power of imagination. With ideas. Then we give them freedom to grow. Thus, we keep broadening our ability to serve our customers. Celanese®

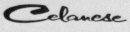

CHEMICAL COMPANY

Celanese

It wasn't so long ago that when you browsed through magazines like *Chemical Week* or *Chemical and Engineering News* there were hundreds of ads that used beakers, flacons or mortars and pestles as attention-getting devices in main illustrations. Too many copywriters of the era seemed to think that chemical engineers, bench chemists and the like could only be attracted to ads that gushed with the paraphernalia of chemistry. Copywriter Bob Moeller of the West, Weir & Bartel agency (later merged with D'Arcy-MacManus & Masius) changed all that with a real offbeat corporate campaign for Celanese Chemical Company. Launched in 1964, the highly impressionistic abstract art technique was designed to project the image of Celanese as a major chemical producer and as a progressive, well-run company with inevitable growth possibilities. Dartnell's *Advertising Manager's Handbook* lauds the campaign and proudly points out that "Celanese ads have constantly been in the top sixth Starch readership surveys. Benchmark studies and follow-up surveys have shown that the company's audience has increased about 10 percent in its recognition of Celanese." Every time reprints were ordered the company was deluged with additional requests, so back on press they'd go to satiate the demand for more copies.

The abstract art/main illustration format was ideally suited and flexible enough so that product ads could be produced along the lines of their corporate "sisters." To this day, Moeller's copy is still being touted by academia. The campaign has been used as graduate course material by Northwestern University and New York University. The fact that Celanese ran the campaign for 10 straight years attests to its greatness as well as its durability.

In 1904, I put all our advertising into Chemical Engineering magazine.

Within a year,
I had cracked the Chemical Industry.

Chemical Engineering

Go through any issue of *Advertising Age, Media Decisions, Adweek* or *Industrial Marketing* and look at the media ads. I'm sure you'll agree that 90% of them are dull, boring, pedantic and ill conceived. Why are industrial publishers—supposedly a bunch of professional communications experts—so terribly bad at producing ads for their own product? Why do most publication-print-ads violate every rule in the book? Most of them show graphs, bar charts and statistics that make you want to yawn and quickly turn the page.

The one brilliant exception is this wonderful campaign McGraw-Hill's *Chemical Engineering* has been running for a number of years. The series of ads has been designed to establish the growth of the chemical process industries and to establish the importance of *Chemical Engineering* as a vital tool that can help advertisers reach that market. This four-color ad (actually it's four consecutive right-hand page ads) proves how effective the short copy approach can be. Each of the pages are dominated by this beautiful, 90-year-old toothless gent straight from the geriatric ward of the Funny Face model agency. Dressed in a baggy morning coat and striped pants and sitting in an antique chair, he says on the *first page,* "In 1904, I put all our advertising into *Chemical Engineering* magazine. Within a year I had cracked the Chemical Industry."

On the *second page,* "In 1920, we were still putting all our advertising into *Chemical Engineering.* By then, I had penetrated fertilizer, explosives, and pulp and paper markets. 1930 took us into petroleum, refining...."

On the *third page,* "And still with just *Chemical Engineering.* In 1940, we grabbed off the paint, fats and oils and rubber products market. In 1950, synthetic fibers.... In 1966, soap...."

On the fourth and *last page,* "Please, *Chemical Engineering,* leave me alone." This is followed by a beautifully written, short copy pitch for the CPI and, quite naturally, *Chemical Engineering.*

The agency was Warner, Bicking & Fenwick. Jack Warner was the writer; Rod Capawana, the art director.

How casting about for cheap parts can turn lift truck repair into a can of worms.

There's nothing wrong with saving money. We recommend it. So we understand why you shop around for lift truck parts. However, if you opt for cheap ones, you could be falling into a very costly trap.

Low price, improper pressure

You've been having trouble with engines heating up. Oil pressure falling off. You've even had some engines with severely scored cylinder walls and pistons or gauled cranks. What's causing it?

Think back. Where did you buy your last batch of oil filters? Are they bypass or full-flow filters? Clark engines are designed to use bypass-type filters. They have an orifice in the plate that builds up resistance to maintain proper oil pressure.

The orifice must be a specific size, or you'll get too much or too little resistance; simply, improper oil pressure. And this function is extremely critical during idle speeds. So where would you rather save money?—Buying cheap, improper oil filters, or buying Clark filters that match engine design, thus saving expensive engine overhaul and excessive down-time?

You'll get the point-almost daily

You've got a real problem. You can't keep ignition points in your trucks. It's costing you a fortune in down-time and maintenance. Your mechanics are having a hard time finding the problem.

Everything in the electrical system checks out; yet, the problem persists.

Better check the coil. If you bought an automotive coil, you're probably not getting enough reserve power. Clark engines are low-speed power plants. Maximum RPM is about 2,350, compared to over 4,000 RPM for an automobile engine. The typical auto coil is a 6 or 8 volt unit that won't generate

enough reserve power to start and keep a Clark industrial engine running properly. Clark engines require 12 volts for a higher firing line—more reserve power. Using lower-voltage automotive coils will cause high surge, burn out points on a regular basis.

These automotive coils can't match the impedance of a Clark coil, so condenser and points will not match the coil. Coil life also is affected, because more current is input. And this high energy in the secondary coil creates heat. So the coil may overheat, and fail.

Investment that pays off

Buying cheap parts is false economy. They can eat up maintenance budgets in a hurry.

Your Clark dealer has the right parts for your lift trucks. And, yes, in some cases they cost more than others. But what you may save in down-time, maintenance and performance, could make them cost far less in the long run.

In addition, Clark dealers have access to the modifications, redesigns and updates that go into Clark trucks. So, they have the parts that match the job applications for your specific trucks. The bottom line? If *you* want to save money on lift truck replacement parts, and keep your maintenance from turning into a can of worms, see your Clark dealer. He has the parts for staying ahead.

For the complete story on the quality-difference of Genuine Clark Parts, and why it's important to use them, write for your free booklet, *"Staying Ahead of Downtime Problems,"* Clark Equipment Company, Inquiry Department, P.O. Box 430, Battle Creek, MI 49016.

CLARK
Industrial Truck Division

Clark Equipment

The J.I. Scott ad agency which produced this great Clark forklift truck parts advertisement used mnemonics to a fare-thee-well, and it sure did the job. Jon Hill, Clark Equipment's ad manager, set a modest 5 percent increase in parts sales as a goal. Instead, the campaign came through with a whopping 14.6 percent increase. Also, 83 percent of Clark's dealers participated in the stocking and promotion of the company's OEM (original equipment manufacturer) truck parts. Jim Scott, the agency prexy, used an incongruous-situation caper to pull off the mnemonical scheme.

You have to agree with Scott that incongruity is one of the greatest tools of advertising because, as he puts it, "it has the power to make a connection between things that matter to your buyer and things that matter to you. The combination of incongruity and familiarity makes for impact and memorability." This double-page four-color bleed spread very graphically (if not nauseatingly) depicts a squiggily can of worms on the left-hand page. The definitive headline "How casting about for cheap parts can turn lift truck repair into a can of worms" runs right across the gutter for solid emphasis and maximum readership. The entire ad (headline, main illustration left page, three-column text right page, along with an inset truck parts photo and a Clark vehicle drawing, cum logo) is surprinted on light beige ink. This holds the whole thing together for more reader impact and total campaign continuity.

For some unknown reason, on an industry-wide basis, forklift truck advertising generally gets superior readership. But this particular Clark ad, and, for that matter, the entire campaign has got to be one of the greatest ever produced in this field.

We are not afraid
to entrust the American people
with unpleasant facts,
foreign ideas, alien philosophies,
and competitive values.

For a nation that is afraid
to let its people judge the truth
and falsehood in an open market
is a nation
that is afraid of its people. John F. Kennedy

artist: Kenneth Josephson

Container Corporation of America

 Great Ideas one of a series

For a reprint suitable for framing write: Container Corporation of America, Dept. SL1-201, 1 First National Plaza, Chicago, Illinois 60603

Container Corporation of America

Whoever would have guessed that you could sell philosophy? Or at least get busy corporate-type publication subscribers to read about it? An *Advertising Age* article once quoted David Ogilvy's prediction at the onset of this 31-year-old "Great Ideas" campaign that it would "soon be consigned to oblivion." And he had every right to so prophesy. Container Corporation of America's series breaks every rule to which experts like Ogilvy adhere. The ads use artist illustrations instead of realistic photography. They are really a potpourri of type, seemingly just thrown together. Often copy runs vertically as well as horizontally, and there's absolutely no semblance of balance. Even type columns are nonexistent. The typographer was obsessed with a "flush-left ragged-right syndrome." The quotations are rather lengthy, pedantic and stuffy. They're difficult to read, understand and assimilate. In spite of all this, the campaign, initiated by CCA's founder, Walter P. Paepke, has been extremely successful in distinguishing them from other packaging companies and in creating a fine awareness of the company. Studies have shown that the campaign has achieved top readership scores as well as indicating a strong corporate awareness. John Massey, director of communications, claims CCA receives between 4,000 and 7,000 requests for reprints after each ad runs. And guess what? David Ogilvy has retracted his initial opinion and now calls the Great Ideas series "the best campaign of corporate advertising that has ever appeared in print." Attaboy, David!

Diamond Shamrock

In 1979, Diamond Shamrock ran an eight-page advertisement in *Chemical Week.* The ad is a study in how fine art direction, good typography, mood photography and narrative copy combine to achieve top readership. *Chemical Week* is, as its title says, a newsweekly trade publication that brings readers up-to-date on current events in the chemicals industry. For some reason, chemical companies—Du Pont, Union Carbide, Witco, Allied Chemical, American Cyanamid and many others—do a bang-up job of producing really outstanding advertisements. So to stand out among these aggressive competitors in a book packed to the nines with top scores, you have to write and lay out exceptional ads.

This Diamond Shamrock eight-pager uses a stark black background with exceptional photographs and captions emblazed on it. And,

even though the photo captions are in reverse (the body isn't), they are eminently readable. One reason they're readable is because the advertiser was smart enough to print the ad himself and supply it to the publications as an insert rather than rely on the sloppy printers often used by today's cost-conscious publications. The cover headline outlines what is to come. "Resourceful in energy, technology, chemicals" is followed by separate pages describing "Energy," "Technology," and "Chemicals." Then this is all very aptly followed by the company's list of products, plants and sales office locations. The copy is written more like a story than a boast-and-claim corporate brochure. It's done in such a subtle way, you as a reader don't realize how you are being led into the fold of Diamond Shamrock's admirers. Successful? Was it ever! The insert Starch scored 88% Remembered Seeing and 64% Remembered Reading. That's the highest score ever recorded in *Chemical Week*. Wow!

Downs Crane & Hoist

Don't knock it because it isn't pretty. It's a 1/9-page advertisement that's been working overtime for Downs Crane & Hoist for each of its 35 years. That's right. This little gem will never win an Andy Award nor will copywriter Robert F. Millar ever be inducted into industrial advertising's Hall of Fame by winning the G.D. Crain Award. Somehow, though, he ought to!

As time goes by, this ad has become more and more productive in the quantity *and* the quality of sales leads produced over that span of years. And, it's been outproducing the average ad in the new equipment type publications it runs in by 50%. The headline isn't a sexy stopper. All it does is label and pinpoint the products Downs wants to sell. The copy sure isn't your basic expository style. All it does is tell you that Downs has all kinds of wheels in such-and-such ranges and sizes. Then it succinctly suggests you "write for catalog." How simple can you be?

That's the answer right there. Keep it simple. When you do, the reader often turns into a buyer. Downs sells through independent distributors, and you can bet they assiduously follow up the leads generated by this hard-working fractional.

New snow-bike glides on skis

Du Pont

One way to come up with a great industrial ad is to follow the age-old formula C + D = I. Translated, this reads *"Continuity plus Dominance equals Impact."* (I believe this little gem was formulated by George Lyon, now prexy of Grey, Conahay and Lyon). And that's just what N.W. Ayer copywriter Fred Kopf did in this six-page, four-color insert for E.I. Du Pont de Nemours & Co. For *continuity,* the engineering plastics people at Du Pont ran this beautifully photographed insert eight times during 1980 in *Machine Design,* a real fat trade publication in which even the most outstanding single-page and double-page, four-color ads can get buried. For *dominance,* the cover shows some fine products made out of Du Pont engineering plastics. The headline, "Du Pont engineering plastics—making some things better, other things possible," really stands out in easy-to-read uppercase and lowercase reverse type. Way down in the lower right-hand corner (also in reverse, but in a much smaller type) is the legend, "6 Pages of News from Du Pont." The fact that the six pages are an insert also helped the ad dominate—mainly because Du Pont printed it on stock of its own choice. The stock is far whiter, chrome coated and (even

of tough foamed ZYTEL ST

Du Pont engineering plastic combines load-bearing strength, impact strength and stiffness.

A motorcycle-ski hybrid, Chrysler Marine's "Sno-Runner" owes much of its lightweight ruggedness to one of the newest nylon resins based on Du Pont's "super tough" technology.

The ski segments on which it glides are injection molded in a structural foam grade of ZYTEL ST, reinforced with 33% glass fibers. These skis must support the entire weight of the bike and an up-to-250lb rider. They have to withstand wear, impact and vibration while providing a secure mounting for the frame and engine.

In addition, the center and rear ski assembly serves as an elongated leaf spring. Here the shock-absorbing qualities of ZYTEL ST cushion both bike and rider.

By using a foamable grade, the designers were able to minimize warpage in the thick sections and increase the size of skis for operational stability, while holding down overall weight. In the rear ski, which supports track and engine, ZYTEL ST offered the relatively low notch sensitivity required in this critical area.

The ultimate choice of the material is perhaps typical of the sophisticated materials selection process in today's designs. Prototype testing soon ruled out most competitive candidates. "What we really wanted," says the chief engineer, "was a nylon with load-bearing strength and flexural modulus close to reinforced ZYTEL and impact strength and toughness approaching ZYTEL ST. We probably got the best of both worlds."

To get the best of the wide world of Du Pont ZYTEL nylon resins for *your* designs, dial Du Pont first. You'll find examples of other Du Pont plastics applications on the following pages. ▶

though it's usually against the rules) somewhat heavier than the run-of-the-book paper *Machine Design* uses to print its editorial and ads on. Inside, case-history copy and crisp color photographs of each product in use add a testimonial-type believability to Du Pont's message. Although the insert is an industrial ad running in a trade publication, the end products are strictly consumer products. Because readers don't expect to see them, this also adds a tremendous amount of dominance to the continuity already established by the year-long media insertion schedule.

Now as far as *impact* is concerned, in five of the eight insertions, the ad ranked number one. It was the highest scoring ad in total readership. In the other three insertions, it ranked number two versus some 150 competing ads in each issue. Hugh B. Horning, Du Pont's ad manager, claims that the "Dial Du Pont First" phone number prominently listed on the back page of the insert was prime generator of some 10,000 prospect phone inquiries. A 1980 tracking survey showed that this single insert was responsible for creating the number one leadership image for Du Pont engineering plastics. Bottom line traceable sales from inquiries resulted in millions of dollars of new business.

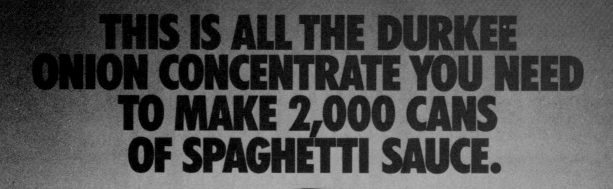

THIS IS ALL THE DURKEE ONION CONCENTRATE YOU NEED TO MAKE 2,000 CANS OF SPAGHETTI SAUCE.

Durkee onion concentrate is as potent a replacer for dehydrated onion powder as you'll find anywhere. And it comes in the most economical form, too.

In liquid form, one pound of Durkee onion concentrate equals one hundred pounds of dehydrated onion powder. Durkee onion concentrate also comes in dry form with intensities from one to two-hundred and fifty fold.

Durkee onion concentrate can be used in gravies. Salad dressings. Sauces. Soups. Just about anything that calls for onion flavor.

What's more, with Durkee onion concentrate you get a continuous flavor level.

And it's industrially sterile.

Of course, the actual amount of Durkee onion concentrate you use will vary according to your recipe and specifications. And the type of soup, sauce or gravy you're making.

But we can promise you this much: nothing will give you more onion flavor than Durkee onion concentrate. Not even (forgive us Mother Nature) an onion.

Call Ed Vining, Manager of Technical Services. Ask him about our garlic concentrate, too. (216) 771-5121.

Durkee
SCM. DURKEE INDUSTRIAL FOODS GROUP
SCM CORPORATION, CLEVELAND, OHIO 44115

Durkee

Twenty-six prominent industry business communicators served as judges for the Business/Professional Advertising Association's 1980 Pro-Comm International Awards. That's a lot of judges, but then they had some 1,300 entries to score, and it took them two days and one evening to do it. It's a Herculean task, believe me, because I did it one year and I'm not looking forward to being asked again. One of the division winners that year was this really outstanding Durkee Industrial Foods ad. It won the top score of all the single-page, four-color ads entered in this category. The man behind it at the Meldrum & Fewsmith agency in Cleveland was Vice President and Creative Director Chris Perry. He was ably helped by Ron Stanger, Durkee's ad manager. The photograph alone was purposely made life-size to make a point. That hand holding the small vial of onion concentrate comes off the page like a real hand. It's not the world's most creative idea for a photograph, but when coupled with the strong headline (and the hand and vial actually point to the headline), the ad comes off sensationally well. You can actually see and hear a Durkee salesperson saying to a customer: "This is all the Durkee onion concentrate you need to make 2,000 cans of spaghetti sauce." The discursive copy that follows really hits the mark: "In liquid form, one pound of Durkee onion concentrate equals one hundred pounds of dehydrated onion powder" and "nothing will give you more onion flavor than Durkee onion concentrate, not even (forgive us Mother Nature) an onion." It's enough to want to make you burp—and buy.

"Oh, go ahead, Clarence, indulge yourself. What if you're ever lost in the North Woods?"

Fascination with multiple-purpose devices is seldom rewarded with truly useful products. We are pleased to offer an exception: the Kodak Ektaprint copier. It does enormously complex jobs involving automatic collating, stacking and stapling. It can also make a single copy with equal efficiency.

Result? Versatile, productive, easily managed output. The Kodak copier does more, all by itself, than a Swiss army of lesser machines all over the place.

May we demonstrate?

Write: Eastman Kodak Company, **CD2356**, Rochester, N.Y. 14650.

Kodak copiers. Everything they do, they do well. And they do everything.

Eastman Kodak

Kodak had a problem in the copier marketplace. Xerox dominated the field for many years because it got to the number one sales position before anybody else and stayed there. Its ad budget was 10 times as big as Kodak's, so quite naturally the Xerox name overshadowed them, particularly where high-volume copier users were concerned. In order to build an awareness that Kodak was a responsible force in the copier business, its agency, J. Walter Thompson Company, came up with this hard-hitting, single-page, black and white cartoon campaign. *New Yorker* cartoonist Charles Saxon was hired to render the main illustrations which were run over a typical *New Yorker* punch line. The humorous line of course became the sweetener which made you hungry to read the copy. The short body text runs about 100 words or so. Not a lot, but just enough to inform the reader that Kodak has the solution to the problems that most frequently face copier users. The campaign was an instant success. At a fraction of the budget, Kodak achieved awareness levels equal to its big-spending competitor. Insofar as readership was concerned, Starch studies showed that this great campaign achieved 29 first-place scores for rank in issue by cost-per-reader. This fantastic Swiss Army knife ad received the highest "recall reading" score of all black and white pages over a five-year period in *Fortune* magazine. Something else!

How many times can this nut be re-used?

ON ANY BOLT OF STANDARD QUALITY, THE NYLON INSERT ELASTIC STOP® NUT PROVIDES DEPENDABLE LOCKING TORQUE

for over 50 on-off cycles

The remarkable wear resistance of the tough nylon collar plus its elastic recovery characteristic make it possible to remove and re-use the standard Elastic Stop nut at least fifty times. This familiar red collar—an integral part of an Elastic Stop nut—grips the entering bolt threads with a perfect fit which dampens impact loads and resists turning under the most severe conditions of vibration and shock. When the nut must be removed for routine maintenance, the nylon collar tends to resume its original shape and, on re-installation, grips the bolt threads as effectively as on the original installation.

Prove it to yourself! Check the coupon for a copy of Recommended Test Procedure for Determining Re-usability.

Re-usability is just one of the advantages of the nylon insert Elastic Stop nut. The constant torque that locks the nut at any position on the bolt: the inertness to gasolines, oils, salt atmospheres, cleaning compounds and common acids: the easy identification on the assembly line or in the field: the one piece construction that simplifies installation and reduces cost—these special features have made the Elastic Stop nut the standard of industry for tough applications.

Elastic Stop nuts are available in thin and regular height hex types in sizes ranging from a watchmaker's 0-80 through 3 inches, also many special shapes to meet your unusual design problems. In standard finishes and materials including carbon and stainless steels, brass, duronze and aluminum.

ELASTIC STOP NUT CORPORATION OF AMERICA ESNA

DOUBLE DEPENDABILITY
The dependability built into every Elastic Stop nut builds itself into the dependability of every product on which it is used.

Elastic Stop Nut Corporation of America
Dept. S31-54, 2330 Vauxhall Road, Union, New Jersey
Please send me the following free fastening information:
☐ Recommended Re-usability Test Procedure
☐ Bulletin No. 5901 showing stop nut design applications on heavy-duty equipment.
☐ ESNA condensed hex nut catalog No. 706.

Name_____
Title_____
Firm_____
Street_____
City_____ Zone___ State___

Elastic Stop Nut

In the early 1950s the name of the game for Elastic Stop Nut Corporation was to generate lots of qualified inquiries. The company sold mostly through independent sales representatives, and these people relied on a continuous flow of sales leads to pay for their groceries. Fasteners are essentially a low-unit-cost, high-volume product with literally thousands of applications going across the entire manufacturing and fabrication spectrum. Although competition was fierce, ESNA had the reputation of being a manufacturer of premium products. The fact that Elastic Stop nuts were precision-made was brought out in every ad the company produced. The "double quality/dependability" theme—use our dependable product and you'll build a dependable product—really helped enhance the ESNA name. Bruce Linck, Elastic Stop Nut's ad manager for approximately 20 years (now retired), had the single philosophy of always going with ads that proved they worked. A number of the real "pullers" like this one were used over and over again without change for dozens of years or longer. Some of them were used in several sizes, depending upon the publication involved. Full-page ads in one or two colors were run in publications like *Product Engineering;* others, in 2/3- and 1/9-page versions in the new equipment books like *Industrial Equipment News* and *New Equipment Digest.* Oddly enough, the fractional pages would do as well or even outdo the full-pagers insofar as inquiry generation was concerned.

Fafnir

"One picture is worth a thousand words" is an overworked aphorism among advertising people. But in the case of this stunning Fafnir Bearings campaign, it really hits the mark. Ted Rozinski, Fafnir's ad manager, had been running two separate ad campaigns. One to reach original equipment manufacturers and the other to hit distributors who took care of the bearing replacement market. He and his agency representative, Dean Wood of Creamer Inc., decided that this not only fractionated the thrust of Fafnir's ad messages, but it was expensive as well. So they came up with this brilliant four-color spread campaign that was designed to reach both audiences with a single message. The idea was to reinforce Fafnir's image as a dominant factor and leader in the bearing business.

The agency must have spent hundreds of hours contacting stock photo houses and screening photographs. I understand they reviewed 20,000 pictures to produce just five ads. But whatever the time involved, it was well worth the effort. The extremely dramatic photographs make the series zing. My favorite is this roller coaster ad which screams at you to read the copy. If you've ever been on a roller coaster you will know what I mean. Short, well-edited copy points out the advantages of consistently attained top-of-book readership scores in all the design engineering and purchasing publications in which they run. And, in 1981, the campaign won an American Business Press O&R Award. But most important of all, it directly influenced a million-dollar buyer to place his order with Fafnir. The buyer admitted he was convinced by this super ad campaign.

WHICH LETTER WOULD YOU OPEN FIRST?

Probably the one that looks the most important. The one that got to you overnight. The Federal Express Overnight Letter.℠

It holds up to 10 pages (2 oz.) of critical correspondence. It's water- and tear-resistant. It goes to any of 13,500† cities. It gets there as fast as we get packages there—absolutely, positively overnight.†

And it costs only $9.50 when you take it to one of our many convenient drop-off locations.

If you want us to pick it up, we will—for the same $9.50 per letter if you're sending two or more Overnight Letters or if we're already picking up a COURIER PAK® or package at the same time. If you're sending just one Overnight Letter, we will pick it up for $18.

Think of it. From now on, all of your important letters can get there overnight, be opened first, and acted upon faster than all the others. For only $9.50.*

Call us at 800-238-5355 or in Tennessee call 800-542-5171 if you have any questions. And if you'd like a free kit of Overnight Letter envelopes and airbills, send in the coupon below.

```
For a free kit of Overnight Letter materials,
send to: Federal Express Corp., P.O. Box 727,
Dept. 250-OL, Memphis, TN 38194
Name_____
Title_____
Company_____
Address_____
City_____State_____Zip___
Phone #_____
Account # (if any)           FNA9
```

*Higher price to pick up only one. †Monday through Friday. Saturday delivery by special request only, and at an additional service charge. Areas served, delivery times, and liability subject to limitations in the Federal Express Service Guide. ©1981 Federal Express Corporation. COURIER PAK® is a registered trademark of Federal Express Corporation.

FEDERAL EXPRESS INTRODUCES THE OVERNIGHT LETTER. ONLY $9.50.*

Federal Express

A headline in a 1982 issue of *Advertising Age* read, "Ally & Gargano rides Express to Clio success." Translated this pithy statement meant that the creative types at this super agency won eight TV Clio awards and that *five* of those were for its Federal Express advertising campaign. One of the five was heralded as the Best of Show for a national campaign. Ally & Gargano's ancillary print campaign has been a staunch supporter of its superb television series. In spades. Of all the corporate and business print ads that have been run by the Federal Express people, all in four-colors (but some in one- and others in two-page size), the best is "Which Letter Would You Open First?". The rhetorical question headline technique often works wonders in advertising, and it did here for sure! No answer is expected to the question because the answer is just too obvious. Michael Tesch, A&G's creative director, and his photographer certainly saw to that! The campaign has been running in *Time, Business Week, Newsweek, Fortune* and the *Wall Street Journal*. This one was researched when it ran in *Fortune* and *Business Week* and its scores exploded! In *Fortune* alone, of the subscribers interviewed after reading an issue, 67% remembered having read enough of it to later identify Federal; and according to this happy advertiser, coupon returns have inundated them with requests for their free kit of overnight letter materials.

"I'm a Mensan, president of my company, and I offer all my key employees Corporate LifeCycle."

She knows how to make points.

Like any employer who knows key people are worth the Corporate LifeCycle investment.

Corporate LifeCycle is a new corporate executive life insurance benefit that gives you and your company a better financial break.

Your company pays all the Corporate LifeCycle premiums with tax-deductible corporate dollars.

And there's little, if any, tax liability for you.

What's more, Corporate LifeCycle keeps on working after most plans stop.

You continue to get lifetime protection without any personal premium cost.

You continue to protect your estate with tax-free dollars.

You continue to protect your family's lifestyle.

And a lot more.

So if you're an executive who wants your family to be well taken care of, or you're an employer who wants to attract top people and keep them, do this. Drop your business card in the mail to:
Corporate LifeCycle
Fireman's Fund American Life Insurance Company,
1600 Los Gamos Rd.,
San Rafael, CA 94911.

We'll send you all the Corporate LifeCycle facts. Then you can show your company how they can make points.

FIREMAN'S FUND AMERICAN LIFE INSURANCE COMPANY
SUBSIDIARY OF AMERICAN EXPRESS COMPANY

Fireman's Fund

Playwright George S. Kaufman once said, "Satire closes Saturday night." Many professional ad writers heartily concur and therefore assiduously avoid print advertisements that use humorous cartoons for their main illustration. Here's a case where these sages are proven wrong. Cunningham & Walsh copywriter Dalton O'Sullivan and his art director Cal Anderson came up with this campaign to help convince top business management that Fireman's Fund's new Corporate Life Cycle life insurance benefit offered executives a perk worth finding out about. Knowing that life insurance has a built-in boredom factor insofar as readership is concerned, this creative team opted to build attention through an assumptive headline and illustration technique that mimicked *New Yorker* cartoons. Although no coupon was used, the tag line at the end of the punchy copy asked interested readers to mail in their business cards to Fireman's Fund headquarters. Once the ads started to appear in the *Wall Street Journal* and other publications, the advertiser was deluged with mailed-in business cards. They also received a great many telephone calls: "Just saw your ad. We're having a board meeting tomorrow and we'd like you to send someone over to explain your program." The first flight of ads were so successful that the client had to cancel the remainder of its media schedule in order to catch up with the steady influx of inquiries.

Sometimes it just doesn't pay for an advertising agency to do good work.

Not for all the tea in China

Unlike the Russians (and their Pepski generation), the Peking regime in China isn't about to encourage the sale of Coca Cola—or any other American soft drinks. At least not now.

Will things go better with Coke in China if and when formal recognition from the U.S. comes? Maybe so. But one thing is certain. With one-fifth of the world's population, and an economy 50% larger than Britain's, China has the long-range interest of many an American marketer who is export-oriented.

In a recent article "Lightbulbs for the Lamps of China?", Forbes takes an illuminating—and highly scrutable look at Mainland China's import/export picture now, and in the not-too-optimistic future.

It's the kind of important, timely editorial coverage that continually attracts the readership of America's key executives. Those at the very top. And those determinedly on their way there.

In fact, Forbes rates first in the measured reading preferences of America's top management. The research firm of Erdos & Morgan made a reconfirming study of this among the corporate officers in 1300 of America's largest companies. The results of this study showed Forbes to be read by more of these top management executives than any other major business or news magazine.

No wonder Forbes was the *only* magazine in its field in 1976 to register a second record-breaking year in a row for advertising page gains. And is the clear winner as the fastest-growing business or news magazine of the past decade, with an advertising page gain of 72%. Compare that, for example, with Business Week—down 28% in that 1966 to 1976 period. Or Fortune—down 26%.

We look forward to new Forbes records in 1977, created by advertisers who select their media on the basis of advertising performance—ours and theirs.

Darn clever, those capitalists, to know that we're just their cup of tea.

FORBES: CAPITALIST TOOL

Forbes

Forbes ran this black and white ad in 1977 in the *New York Times, Advertising Age, Madison Avenue* and of all places the *Wall Street Journal.* And Doremus & Co., the New York ad agency, ought to give art director Tom Van Steenbergh a big fat bonus just for casting the photograph. The sullen Chinese model is the single most important element in the entire ad. Talk about perfect weight and focus. This is a superb example of trapping a reader's eye and dragging him bodily to the page. That stubborn look of disapproval and rejection really gives the audience some deep insight into the Oriental mind. Combine this with a very descriptive and meaningful headline and a piece of long but great body text, and you've got a formula that insures instant readership. Let's not forget *Forbes'* provoking slogan which is still being used today. *"Forbes:* Capitalist Tool" really tells advertisers all about the value of using this very prestigious medium. Copy was written by Gene Gramm, who also is responsible for *Forbes'* excellent four-color campaign. Malcolm should be very proud of this one. Let's hope he treats these fine agency craftsmen to a free trip aboard his fabulous boat, the Highlander.

Foxboro

The reason Foxboro's "From Pigs to Purses" is so good is because it's so different. Who would ever expect to meet a pig head-on in a very technical industrial trade publication? Forty-three years ago, Foxboro ran this two-color insert on a heavy stock (which in itself helped draw readers, unfairly perhaps, because magazines always fall open to bulky pages). The copy of the "pig" side of the insert is written in a consumer style with a bit of whimsey. Its short breezy captions and simplistic line drawings readily explain the wide application of Foxboro temperature controls within the meat packing industry. Foxboro's sow's ear obviously is going to put lots of savings into its customers' silk purses. On the flip side of the insert, the copy is tradey and as technical as an industrial bulletin. And that's the beauty of the ad. The pig stops you and gets your attention, and then the catalog gives you all the technical information you need to know on temperature control instruments. The credit for this "super sow" goes to Henry Silldorf of the old G.M. Basford agency and Ed Lawson, the former ad manager of Foxboro.

GAF

GAF Corporation was saddled with a foreign image. Although it was American in origin and in manufacturing, the company was once owned by a German conglomerate, and to many the acronym GAF spelled out "German Air Force" rather than "General Aniline and Film." The people they wanted to sell—professional photographers, graphic arts types, and medical and industrial radiographers—perceived the company as foreign and were constantly worried about a continuous and ready supply source. To the rescue came the inseparable but talented creative team of Allan Shaw (art) and Ken Todd (copy), who at the time were working at the now-defunct Michel Cather agency. They still handle the GAF account, but now it's at their own shop. (See, it sometimes pays to be truly creative.)

Shaw and Todd kicked off the campaign with a four-page insert. The first page read in

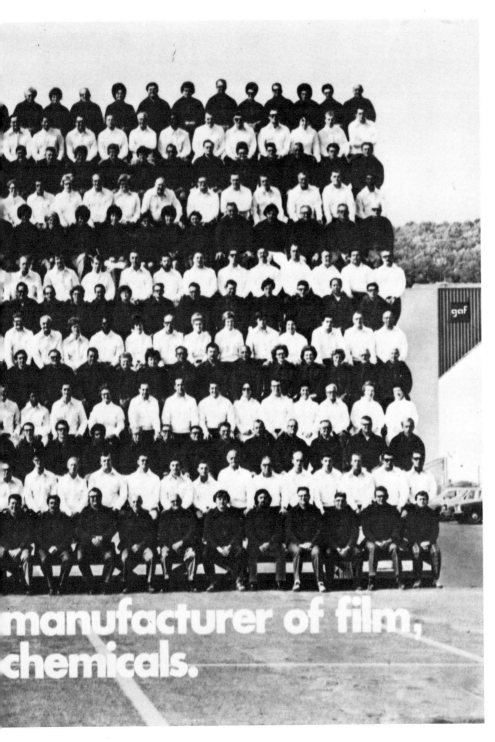

big loud type, "Some people think GAF is a foreign photographic company. We're not...." The inside color spread which you see here was a living American flag comprised of 332 GAF employees, photographed in the parking lot of a GAF plant in Binghamton, N.Y. (The photo is actually five separate shots artfully pieced together.) The GAF people wore blue shirts for the stars background and red or white shirts for the 13 rows in the flag. What a photo session that had to be! Of course, page four reinforced the message with a historical capsule about the company and its capabilities. The American flag artwork was then used as a graphic element in every product ad for the next two years. And did it ever work! A followup study showed that the image had been exactly reversed; instead of 90% of its prospects and customers recognizing GAF as a foreign entity, 90% now recognized it as truly American.

Gates

Gates Rubber Company in Denver makes a lot of ingenious rubber products. The company also makes a lot of really great ads with its advertising agency, Broyles, Allebaugh & Davis Inc. This super ad is part of a campaign that's been running in *Industry Week,* and it must be doing something right because it's won two B/PAA Pro-Comm awards—one in 1980 and one in 1981. This double-page, four-color spread is another excellent example of how the proper use of colors in photography can attain readership. The stark all-white motorcycle rider certainly contrasts with the

Chains and motorcycles have been linked since the first motorcycle rolled out of the inventor's garage.

We have just broken that link. Harley-Davidson is now using Gates newly developed toothed Poly Chain® Belts for the primary and secondary drives on their Sturgis model.

That means a quieter primary drive. And a secondary drive that never needs lubrication and rarely requires adjustment. Two messy and annoying chores that are now a thing of the past.

Industry has turned to Gates for over 70 years to provide solutions to difficult problems, and proven performance in efficient, reliable rubber products. The Gates Rubber Company, 999 S. Broadway, Denver, CO 80217.

RubberEngineering.
The payoff is performance.

motorcylists you see driving along the highways today. He's so pristine-looking he makes you conjure up a white knight riding lickety-split into battle to avenge some sultry Lady Marian. The black motorcycle of course is his faithful steed. And, guess what? It sports Gates newly-developed toothed Poly Chain Belts for the primary drives; and they make motorcycling cleaner, quieter and a lot less work according to the headline. In seconds, you the reader are drawn into the copy. And really that's what good advertising is all about. Larry Hennesy, Gates' Industrial advertising manager, ought to get a big fat raise for this one.

A "small businessman" tells...

"WHY I GOT MAD AND WROTE MY CONGRESSMAN"

"... I am president of an extremely small and young company, the Elano Corp., located in a rural community near a town most people have never heard of, Xenia, Ohio. Our production facilities are in the basement of a public garage. Our office is on the ground floor where we share the available space with the town fire department and a new car agency and garage.

"In our basement facilities we turn out small parts for the jet engines General Electric is building for the Air Force. We're turning out a lot of them and we're confident that if we continue to produce quality products that we'll very likely be called upon to expand our output. We're on the way up, not out, because of our special skills and engineering background, and because of the dependence of big business like General Electric upon our vital contribution.

"And to be perfectly frank, we are equally dependent upon big business for several reasons. First and most obvious, because big business serves as an ideal outlet for manufacturers of component parts; second the extensive purchasing experience of big business can prove invaluable in obtaining hard-to-get materials to the relatively inexperienced small businessman; third because we will benefit from, and can capitalize on laboratory research results made available to us by big business, research we can neither afford nor spare the time for.

"This may sound like a one-way street—that the small businessman is getting the better deal. In our case, that definitely is not so. We brought to General Electric a background of proven ability to produce tube assemblies, a solid development engineering know-how gained from job experience, and the will to produce a first class product.

"We're pulling ourselves up by our own bootstraps, I know, but it's a fine old American custom. Now, we're gambling, but with good odds, that our ability, faith in the basic rightness of free enterprise, combined with a small talent for parlaying a dollar will add up to success for the Elano Corp.

"That's why I get good and mad when I read irresponsible charges that small business is being pushed out of the defense picture. We've got good proof that it isn't—right here in Xenia, Ohio."

ERV NUTTER, president of the Elano Corp., wrote his congressman the letter quoted at the right. The knowledge and skill of Mr. Nutter and his employees has tripled Elano's business in less than a year. The Armed Forces encourage subcontracting because it broadens the defense-production base.

Footnote: In a speech to the House of Representatives, after reading the Nutter letter, Representative Clarence J. Brown (left) added: "We need both big business and little business, industries employing thousands of men, and modest concerns employing only a few dozen—all working together to build such a strong defense as will guarantee our security."

General Electric

One of the most famous of the old corporate campaign genre is this General Electric effort that ran a quarter of a century ago. The Korean War was on and G.E. was getting fat on jet engine contracts. Leading questions were being asked in Washington about war profiteering. Sound familiar? Big business always seems to get blamed just at the time our country needs its help most. General Electric and its apparatus division ad manager, Stan Smith, figured the problem could be answered by informing its decriers that G.E. was only a prime contractor. The company was actually subcontracting much of the work out to lots of smaller companies, all of whom were simply wonderful American people who ate apple pie, believed in mother, God and the free enterprise system.

Stan got together with G.M. Basford copywriter Ed Hatcher and art director Pat Pataky. Together they came up with this black and white bleed spread campaign that

prominently depicts small business people who were G.E. subcontractors. "Why I Got Mad and Wrote My Congressman" features a Xenia, Ohio (how close to God can you get?) businessman, his pretty peaches-and-cream-complexioned assistant treasurer, a congressman, the Xenia high school principal and even the local grocer. Each attractive photograph is captioned and the entire layout emulates the *Life* magazine format (which was extremely popular at that time). The fact that the ads ran in *Life* helped readership tremendously.

Each year the ad series was bound into a booklet entitled *Dedicated to Defense*. At the back of the booklet was a beautiful color shot of Ronald Reagan, the host of the "General Electric Theatre," who plugged the print ads on the air. Every year G.E. took the ad booklets down to Washington and merchandised the campaign around the halls of Congress. The fact that lots of congressmen wrote General Electric expressing their appreciation was proof that America was indeed "putting their confidence in General Electric."

THEY "SAVE EVERYTHING BUT THE SQUEAL"—WITH RUBBER

A Typical Example of Goodrich product development

PACKING plants found out, a few years ago, how to salvage the hair from the hog, and so "save everything but the squeal." Giant rubber scrapers churn, and scrape the hair from the carcass. But 200-pound hogs soon wear out and break off the scrapers—hair becomes expensive.

Then a manufacturer of machinery had an idea and came to Goodrich with it. We helped him work out a scraper of a different rubber compound and shape, that stands up under the weight and impact of the hog and resists the chemicals in the scraping bath. These Goodrich scrapers will clean 50 to 100 per cent more hogs than the old type.

Costs cut substantially! Remarkable? New Goodrich developments are doing it nearly every day. Rubber, as Goodrich can now compound it, will—

—flex indefinitely without breaking ... and Goodrich transmission belts set new performance records.

—withstand abrasion ... and Goodrich gravel chutes, ball mill linings and a hundred other applications outlast steel 10 to 1.

—resist chemicals, oil, time itself ... and Goodrich-lined tanks and pipes drastically reduce pickling, plating, chemical handling costs; Goodrich hose lasts longer; Goodrich gaskets form life-long seals.

—adhere to metal ... and industry seizes upon abrasion shoes for airplanes, vibration dampeners for countless products, rubber-lined tanks and tank-cars.

Goodrich is always busy with development work on rubber. No product is too "staple" to share in this improvement which extends to hundreds of items made by The B. F. Goodrich Company, Mechanical Rubber Goods Division, Akron, Ohio.

Goodrich
ALL *products problems* IN RUBBER

B.F. Goodrich

When in doubt, use Ayer No. 1 layout (or "A" layout as it's aptly referred to most of the time). All you do is use a big, attractive, attention-getting photograph. Position it over a provocative, user-benefit headline. Then stick the headline on top of some hard-hitting case-history copy and spot in a nice logotype in the lower right-hand corner. Result? An advertisement like this fine B.F. Goodrich single-pager that started running in 1934. The campaign was created by Kenneth W. Akers, who was later to become president of the Griswold-Eshleman ad agency, and H.E. Van Patten, one-time ad manager of B.F. Goodrich Industrial Products. The fact that these ads ran in this format for more than 25 years really attests to its ability to generate results. The time-tested format (combined with B.F. Goodrich's insistence on buying page-one positions in trade publications like *Coal Age, Engineering and Mining Journal,* and business and newsweeklies like *Time, Business Week,* and *Newsweek)* insured readership scores that hit the ceiling year after year. Remember, if you're afflicted with a type A personality, try using type A layout. It will help you produce great ads. Believe me.

This Goodyear steel cable belt averages 92,000 tons of copper ore every day.

This unique 60" mine haulage system attracts mining engineers and students from all over the world. The minus-6-inch ore belt system is over 3½ miles long, and at one point it climbs out of the pit at a 16° incline.

The service is tough, demanding the quality of Flexsteel® steel cable conveyor belting, which is engineered to provide over 3100 pounds per inch of width of sure strength.

The conveyor belt system has a 6400 ton per hour capacity and averages 92,000 tons per day. It's fed from the primary crusher by 33 trucks per hour, 300 per shift. Every 2 minutes, a 150-ton hauler dumps its load into the hopper.

The conveyor system runs 24 hours a day, 7 days a week, except 8 hours on Thursdays for scheduled maintenance. During scheduled downtime, two crews of 11 to 12 men are able to maintain this overland system as well as an additional 7,000 foot, 72-inch wide waste and stacker system which also uses Goodyear Flexsteel belting.

If you're looking for ways to cut the cost of conveying materials, talk to a Goodyear distributor, or write Goodyear, Box 52, Akron, OH 44309.

We know how to help.

GOOD/YEAR
INDUSTRIAL PRODUCTS

Goodyear

This excellent Goodyear advertisement reeks of power mainly because of its powerful headline. The 92,000 tons of copper quickly translates into an astounding 1,840,000 pounds. Multiply this by 360 days and you've got a steel cable belt that moves more than one-half billion pounds of copper ore a *single* year! That's a lot of copper, and that's why this ad blows your mind. Whether you have the responsibility for buying flexible steel belting conveyer systems or not, you still have to be impressed with this Goodyear statement of fact. The four-color photograph also has a tremendous amount of visual magnetism and impact. For some hypnotic reason, when you have a photograph that starts from infinity and stretches right up to the viewers' eyeball, you have a surefire method of attracting page-turners. I guess this third-dimensional effect somehow motivates the reader to stop for a moment and become personally involved in the picture. If you are ever on a "shoot" and you have an opportunity to photograph to infinity, by all means take advantage of it. Who knows, you may come up with another winner like this one.

It is not for us to spin the thread or weave the fabric; we do not make film for the world's entertainment, or colorful plastics, paints, lacquers, and other finishes. We do no mining, quarrying, or construction. Not one pound of paper do we sell. Providing housing, clothing, transportation, communication, entertainment, and adornment for peoples throughout the world—all this, and more, is our customers' job. But—

THIS IS OUR JOB

To take the tiny fuzz from the cotton seed; to purify it, process it, and give it utility. To harvest neglected stumps from cut-over southern pine lands; to remove valuable resins from them, producing rosin, turpentine, pine oil, and an ever-widening cycle of new and useful terpene chemicals. To transform pure cellulose by the magic of chemistry into various cellulose products with a wide variety of properties to fit the needs of diverse industries. To associate nitrogen, carbon, oxygen, and hydrogen in many and intricate combinations, harnessing the tremendous power of explosives to do man's heavy work. To make sizes, coatings, and numerous other chemicals used in all kinds of paper. To study the needs of our customers; to understand as far as possible their problems and their activities, so that we may aid them to use our products effectively. To do all of these things to the best of our ability—this is our job.

HERCULES POWDER COMPANY
Incorporated
WILMINGTON, DELAWARE

WRITE FOR A COPY OF THE BOOKLET DESCRIBING HERCULES PRODUCTS

IN-41

Hercules

"This Is Our Job" is an all-type ad that does just one thing. It informs the reader just what Hercules Powder Company does. It is a corporate-conscience ad in its most simplistic form. The first paragraph is set in a small, 8-point italicized Roman typeface and it starts out, "It is not for us to spin the thread or weave the fabric; we do not make film." This is the device that gets your attention. It's probably the first time anybody ever sat down and wrote the introduction to a corporate advertisement that informs the reader what the company *does not* do. The intro then finishes up by saying that "all this, and more, is our customers' job. But—." The "But—" segues into the headline, "This Is Our Job". The main body text then explains in short, concise words and sentences (there are very few multisyllable words) what Hercules does, and, in so doing, how it helps Hercules' customers. This fine piece of copy was the result of a troika headed by Theodore Marvin, ad manager; Monty Budd, assistant ad manager; and E.I. La Beaume, a vice president of Donahue & Coe, Inc., Hercules' New York ad agency. According to Marvin in a 1939 letter to the editor of *Industrial Marketing,* the copy ran in financial papers, general newspapers and business publications. It also was used as a direct mail piece to customers, appeared in Hercules' employee publication and was used by the president in some of his speeches. It all goes to prove the efficacy of glorious black and white.

Hewlett-Packard

The industrial ad business owes a debt of gratitude to Alan Bugbee, a district sales manager of Penton/IPC Publishing Co. Alan has represented *Machine Design*, one of his company's flagship magazines, for more than 20 years. From 1959, Alan has studied the Ad Gage readership scores of over 100,000 individual advertisements! The man's quest for readership knowledge is insatiable. At the drop of a hat he'll jump in his company car and drive half a day just to meet someone and discuss his tracking studies of great ads. Believe it or not, he carries his research data in a hand truck; it's too voluminous to carry otherwise. Alan's the gent who touted me on to this fabulous Hewlett-Packard spread that ran in *Machine Design* in 1973.

"Increased power at no increase in size or price" is an innocuous looking ad at first glance. Not exactly what you would suspect would qualify for "100 Greatest" selection by any means.

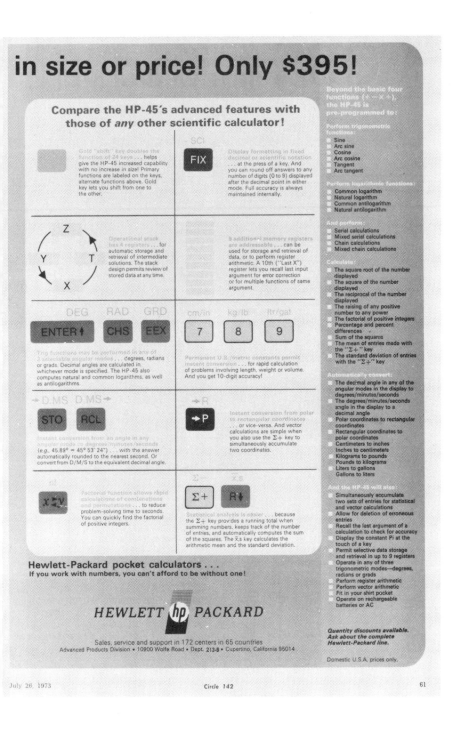

Even though the hand-size pocket calculator was outlandishly priced, even by today's standards, the ad was a tremendous success. Instead of being turned off at the $395 cost, readers flocked to those two heavily laden, copy-filled pages for more information. Their voracious engineering minds wanted hard, detailed facts on how Hewlett-Packard's new calculator could help them in their jobs. And the copywriter did an impressive job in passing along the word. Obviously the product was a real winner, unmatched at that time by any others available on the market. One glance at the face of the calculator showed the engineering audience that Hewlett-Packard's calculator had far more capabilities than most others. This superb combination of product and message turned the trick. The ad earned the top Ad Gage rating ever recorded in *Machine Design*. It's 84% rating is the highest readership mark ever received. Today it is still the number one ad of 100,000 studied in the past 23 years.

Hoffman-LaRoche

Pardon the pun but this is a fine Roche Fine Chemicals ad. Actually it's more than that. It's a great ad! Written by Bob Singer when he was at Marsteller, with art direction by Hank Grapper, the ad has a stunning photograph that has to be the absolute cynosure of every magazine it appears in. A mother breast-feeding her baby conjures up very personal emotions. As a reader you almost feel as though you are invading their privacy just by viewing this special act of love. And that's

crease as consumption of polyunsaturated fats increases. An important sales point, considering the trend towards diets high in polyunsaturates.

Add to the nutritional and marketing effectiveness of your product. Ask us about the different Roche forms of liquid and dry vitamin E, including a new directly-compressible Vitamin E Acetate 50% SD. Marketing Manager, Pharmaceuticals, Roche Fine Chemicals Division, Hoffmann-La Roche Inc., Nutley, New Jersey 07110.

The **vitamin producers**

exactly what attracts and drives you right to the advertising message. "Are you delivering as much vitamin E as the original manufacturer" has an emotional appeal that literally grabs you by the scruff of the neck and pushes you into the copy. For those of you who hate headlines that ask a question, I have a question. Would this ad have been any more effective if the headline had been written as a positive statement? Rhetorically speaking, the answer should be a resounding no.

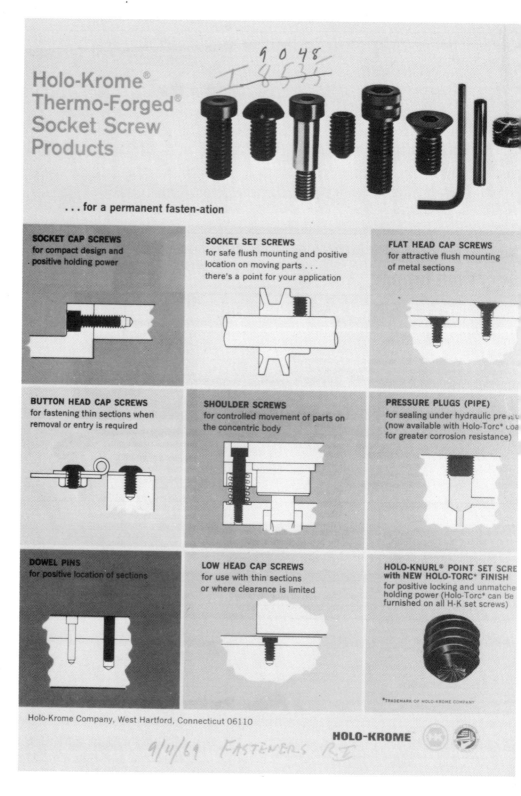

Holo-Krome

Our consumer ad counterparts would retch at the thought of putting this two-color Holo-Krome spread in their proof books. In the first place, it's "catalogy-looking." It belongs in a yearbook or trade directory. The kind of stuff that runs in *Thomas Register*. The banal headline, "Holo-Krome Thermo-Forged Socket Screw Products," reads more like the title of a piece of sales literature, or the heading for a section in the classified telephone book. Dullsville, indeed. But think of yourself as a design engineer who is searching for new and better positive-locking fasteners to use in a new product. You're thumbing through this industrial magazine and all of a sudden you see a Mondrian-like layout which shows nine differ-

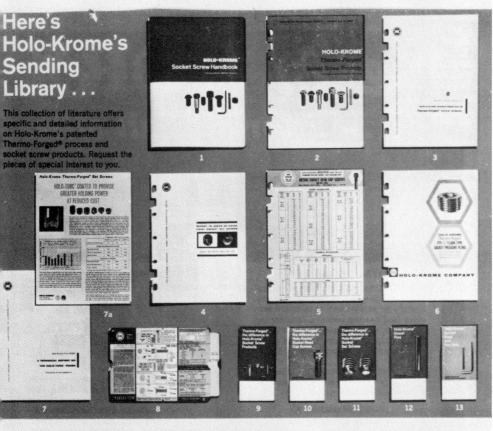

ent but beautiful new fasteners (they're beautiful to you because you need to know about them to do your job). Your grateful eyes then scan the rest of the ad and you see 13 booklets that are available from Holo-Krome's sending library, completely free of charge. To get every one of them, you only have to circle a number in the publication's handy Bingo card. Research told Holo-Krome that ads which offer free literature and show illustrations of that free literature achieve top readership and get scads of requests. That's true because we know the easier you make it for a responder to respond, the more responses you get. In just one insertion, in just one magazine, this splendid effort brought in 9,048 inquiries. Bravo.

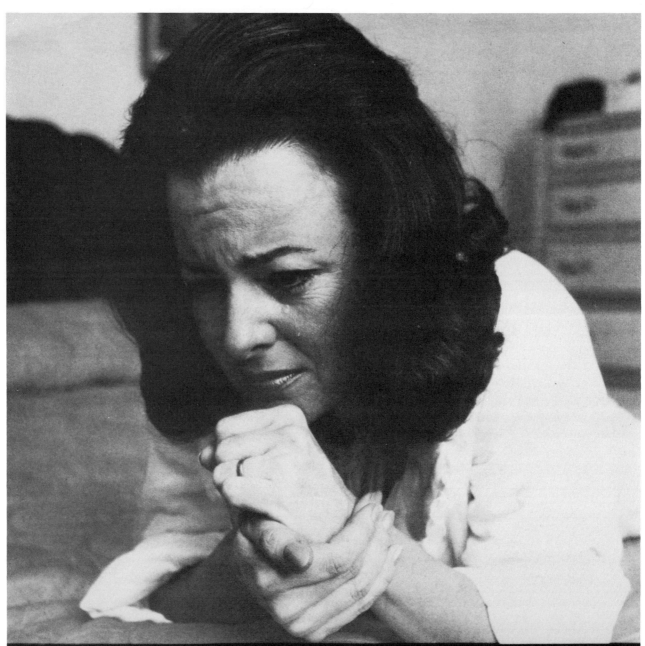

Homequity/ Homerica

Homequity/Homerica was in the corporate home relocation business long before anyone else. Its communications objective was to get top management from as many of the Fortune 500 companies as possible to listen to a Homequity sales presentation. This ad was to be the first in a series that ran 15 years ago in *Duns Review, Personnel Management* and other business trade publications that catered to personnel directors as well as corporate financial executives. Conceived and written by Fergus O'Daly, Homequity's agency's creative director, the ad is a classic in the industry. Because Homequity was deluged with phone calls and requests for appointments, the company had to cut its ad budget in order to pay for the hiring of more salespeople to make the necessary presentations. The photograph of the lady model with tears in her eyes and a perfectly distraught look was a superb way of enticing prospects into reading first the headline, then the body copy. John Huggins, at that time Homequity's president, claimed he didn't have to run any additional ads because this one alone kept his enlarged sales force busy for three full years.

Honeywell

It takes guts to come up with a new, offbeat idea for an ad campaign and then run with it. It takes guts on the agency's part as well as the advertiser's. Who would ever have thought that animal sculptures made out of computer components—lots of wire, resistors, and transistors—would attain good readership scores? Honeywell's Electronic Data Processing people used this extraordinarily outré format in virtually every one of its computer ads for over 14 years. The sculptures literally have become a Honeywell computer trademark. According to Morris Dettman, director of advertising, thousands of ad posters suitable for framing have been sold at cost to customers and prospects, and they're still seen hanging in executive offices and computer rooms all over the world. "There's a little bit of chicken in all of us" is a classic mainly because, without mentioning anybody by name, it's a direct slap at IBM. When it ran, the industry was (and still is) dominated by IBM. People bought IBM because it was a "safe" buy. No buyer would get into trouble by purchasing the number one selling computer in the country. This ad tried to change all that by making you stop and think. Who wants to be called chicken? A lot of prospects with decision-making powers were turned away from IBM as their sole source by this single piece of copy. One high-level executive who worked for a competitor admitted this ad was the key factor in his decision to quit his job and go to work for Honeywell. Produced by BBDO, it was considered a major factor in making Honeywell the world's second-best-known computer company. *Business Week* very seldom gives out awards for advertising, but it gave Honeywell an Advertising Achievement Award for the super readership scores this campaign attained from 1964 to 1967. The series achieved Starch scores of a Noted rating of first in its category 12 out of 14 times during 1967. It also received a Noted ranking of second in its category 25 out of 27 times in 1966-67. It ranked in the top five of all advertisements 90% of the time during 1966-67. Now that's a lot of Starch!

SCIENCE/SCOPE

Diverse educational and training programs help Hughes employees stretch their talents and meet the ever-escalating challenges of constantly changing technologies. Last year Hughes awarded 432 fellowships and 92 scholarships to employees earning masters, engineer, doctoral, and MBA degrees. Also, more than 1,000 employees enrolled in one of the 186 graduate-level engineering and applied science courses taught at company facilities by Hughes professionals. To learn more about opportunities at Hughes, look for our fellowship and college advertisements in forthcoming publications.

An optical chip the size of a stick of chewing gum can do the job of conventional electronics equipment the size of a two-drawer file cabinet in analyzing and identifying microwave frequencies. The chip is called an optical planar waveguide and is part of a larger device known as an integrated optical spectrum analyzer (IOSA). The IOSA uses a beam from a tiny semiconductor laser to separate a broadband microwave signal into as many as 100 individual frequencies. A key feature of the planar waveguide is two concave lenses ground into the chip's surface. The first lens collimates the laser light so it travels correctly through the microwave acoustic signal, which bends the beam. The second lens focuses the bent beam into one or more of 100 charge-coupled detectors. Hughes developed the IOSA for the U.S. Air Force for microwave signal processing.

A prototype of the system that will serve as radar and radio for NASA's Space Shuttle has met its scheduled completion date and is undergoing tests. As a radar, the system will allow astronauts to rendezvous with orbiting satellites in order to repair or retrieve them. It also can track any payloads released from the Shuttle. As a radio, the system will link with the Tracking and Data Relay Satellite System to let astronauts communicate with stations on earth. Hughes delivered the Ku-band integrated radar and communications system, as it is called, to Rockwell International, builder of the Space Shuttle.

If you're pursuing a master's, engineer, or doctoral degree in engineering (electrical, electronics, systems, mechanical), computer science, applied mathematics, or physics, you might quality for one of 100 fellowships Hughes will award this year. Recipents will have all academic expenses paid and will receive an educational stipend, summer employment, professional salary, and employee benefits. More than 4,500 have received Hughes fellowships. For more information write: Hughes Aircraft Company, Fellowship Office, Bldg. 6/C122/SS, Culver City, CA 90230. Equal opportunity employer.

Expanding the use of laser surgery in dentistry, neurosurgery, ophthalmology, and urology may be one benefit of a new Hughes optical fiber. The fiber is made of thallium bromo-iodide, a polycrystalline substance. Unlike an ordinary glass fiber, it can transmit several watts of infrared laser power. Because doctors could use the fiber to direct a laser beam even inside the body, it may one day replace the cumbersome mechanical mirror arrangement now used in infrared laser surgery. Other potential uses are for laser cutting and drilling, as passive detectors in military infrared systems, and for transmitting data and voices.

Creating a new world with electronics

HUGHES

HUGHES AIRCRAFT COMPANY
CULVER CITY, CALIFORNIA 90230

(213) 670-1515 EXTENSION 5964

Hughes Aircraft

How do you build the image of a company that has the word *aircraft* in its name but doesn't build aircraft? And how can you come up with an exciting, vital campaign about a nonconsumer company that is essentially involved in high-tech, egghead stuff like the research, development and production of electronics? Hughes Aircraft asked its agency, Foote, Cone, & Belding, these questions in 1966 and FC&B came up with a choice of 10 different basic approaches. One of these was this editorial-style newsletter called "Science/Scope." And that's the one that Hughes was smart enough to go with. We all know the pulling power of newsletters bound into magazines. When done right, they really work. They emulate editorials and therefore pick up a lot of credibility on the way. "Science/Scope" is a leader of the genre. Its expository writing is ideally suited to the business and technical audiences to which it is directed. There are four different versions that are expressly written to conform to the reading preferences of the four types of publications in which "Science/Scope" appears, i.e., newsweekly, international, military and technical. Each of the full-page ads carry five brief typewritten descriptions of Hughes' products, activities and milestones surprinted over a pleasing 4-A's yellow. Over the years the typewriter type and the yellow color have so much become the company's eyepatch that it's easy for "Science/Scope" aficionados to thumb through a magazine and find their favorite ad. Cindy Baker, Hughes' manager of corporate advertising, proudly points out that "Science/Scope" has received 37 first-place recognition awards since its inception 15 years ago.

IBM

About 30 years ago the hot creative ad agency shops got into realism. Rosser Reeves of Ted Bates wrote a book on it. Bill Bernbach of Doyle Dane Bernbach and a passel of other creative-types pushed for stark realism in photography at the expense of art illustrations. All of a sudden the great commercial artists of the time, the Boris Artzybasheffs, Norman Rockwells and John Falters, were finding it difficult to find work at ad agencies. Photography was king. Today, art is creeping back in. More art directors are tending to turn away from photography and are at last considering the use of art. A fine example is this four-color IBM corporate spread which appeared in *Time* national. Lord, Geller, Federico, Einstein is the IBM agency that produced this one which obeys all the rules on how to prepare a "greatest" ad. The type is set in serif face. The columns are less than 55 characters wide.

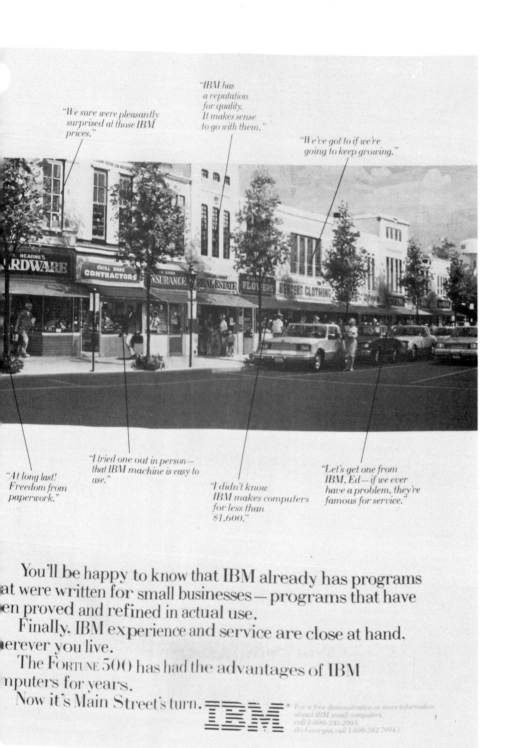

There is no surprinting of text on pictorial matter. There is no reverse type, and everything, even captions, is set in larger than 10-point type. The main illustration just *had* to be artwork. Where could a photographer have found a location that offered a Main Street with so many beautiful store fronts that all required computers inside? The hometown feeling the artist rendered makes you want to window-shop, and naturally once you've stopped you have to read the body text. The exploded capitons for each store also draws you right into the middle of this fine ad. You bet the agency's production budget came in far less than what a four-color chrome would have cost. Add the location costs, the fees for 30 to 50 models, a day's shooting for a photographer, a stylist and two or three assistants; and you've got a bundle of out-of-pocket production costs. It was definitely much easier and far more effective to go the art route.

INA

INA has been in business since 1792 so it's got to be doing something right. One of the things it has had going for it is this hard-hitting corporate campaign which has been hitting stratospheric readership scores in *Business Week,* issue after issue. "New Pension Option for Smaller Firms" is one of the top inquiry-drawers of all time for INA. Even though it's hidden in the last paragraph of the extra-long body copy, the request for more information is all that's required to get people to write in (INA doesn't believe in using coupons). This tends to make you believe that all of that copy is being read. It's a tribute to the copywriter because it takes six minutes to read the ad in its entirety. And believe me, that's a long time to ask a reader to hold still for an ad message. Imagine if you had to sit through six-minute

...ns for Smaller Firms

Some recent developments make it easier—and less costly—for small firms to offer pension plans to their employees.

One is the increased use of money purchase plans which give employers better control of their costs, by defining their contributions rather than the benefits. Another is the growing use of profit-sharing programs which permit companies to add to pension funds in profitable years—and reduce or omit contributions when profits decline.

Thrift plans, where both employer and employee contribute to the fund, also are becoming available for smaller firms. All three—money purchase, profit-sharing, and thrift plans—share a common virtue: employees can watch their funds increasing year by year.

Finally, the 1981 Economic Recovery Tax Act may help both employers and employees. After January 1st, 1982, employees who voluntarily contribute to qualified retirement plans can take these payments—up to $2,000—as income tax deductions.

This more favorable tax treatment may enable an employer to establish a retirement program with much lower company contributions than would have been acceptable before. Voluntary employee-paid pension plans may become a more likely supplement to existing corporate plans.

Simplifying the process

Insurance companies, among others, offer methods that help small businesses reduce the technical work and cost of establishing plans. "Ready-made" prototype plans cut required legal work, ease burdens of ERISA compliance and reduce the need for specialized outside help. Insurance companies also provide a wide variety of services to help with plan design, record keeping, actuarial and administrative work, as well as employee communication.

Some new insurance company plans feature more flexible investment options for growing businesses. One example is the INA-Flex Group Annuity Contract from the Life Insurance Company of North America, an INA company. It provides a wide range of choices in a single package. Companies whose priority is security can put all or a portion of their funds into an account with a guaranteed floor rate of interest. To accommodate other investment objectives, INA-Flex offers two separate pooled accounts, a money-market account and a growth-oriented common stock account. Professional investment managers carefully supervise these investment vehicles.

Comprehensive services

Problems as complex as pension plan development require a coordinated response. The Life Insurance Company of North America's comprehensive approach combines administrative skills with sophisticated investment planning to develop a program expressly tailored to the needs of smaller businesses.

In life and group insurance, property and casualty insurance and risk management services, health care and investment management, INA and its affiliated companies offer a unique combination of products and services to business and industry worldwide.

Founded as the Insurance Company of North America in 1792 in Independence Hall, Philadelphia, today INA Corporation is a diverse organization with an international network of insurance, financial, and health care interests.

For an informative booklet on pensions for small businesses write INA, 1604 Arch Street, Dept. R, Philadelphia, PA 19101.

Counseling the mighty

The largest pension funds—just like some of the smallest—look to outside counselors for investment guidance. Insurance companies are frequently the choice. The Life Insurance Company of North America, for instance, can help growing firms with plans like INA-Flex, while medium and larger companies can seek more specialized advice from INA Capital Management, a subsidiary devoted to investment counseling. The subsidiary's strength: the close attention paid to each client's unique investment objectives.

INA
The Professionals

---

television commercials? INA's ad agency, Geer Du Bois of New York, is responsible for this effort which uses Peter Farago's adroit art direction with dexterity. What probably makes the ad stand out is the *limited* use of four-color artwork on each page of the spread to help tease the reader into the copy. The use of a double-column box replete with an exciting subhead is also a smart device that helps break up the monotony of the lengthy text. The continuous top readership of the campaign is a tribute to the copywriting ability of John Jackson, who was responsible for the ad copy from 1976 through 1980. He makes it easy for the reader to understand the complex jargon of the insurance industry. His short, staccato sentences and the use of words with few syllables are all part of it.

Inland Steel is here... with great expectations

The future is exciting, challenging, and Inland Steel is already at work to meet it. The latest 5-year expansion program costing 430 million dollars increased annual steelmaking capacity to more than 6,700,000 tons, added an automated slabbing mill, new cold rolling mills, a new iron ore mine, a new tin mill, and a giant ore freighter. And already another multi-million dollar expansion has begun with the construction of a completely new 80" hot strip mill—the most modern computer-controlled, fully automated mill of its kind. Two basic oxygen converters capable of producing more than two million tons of steel annually are soon to go into operation. New blooming and billet mills, new welded-beam equipment, new iron ore pelletizing

Inland Steel

You'd never know judging by the photograph that this beauty is an advertisement for a steel company. I guess that's why the copywriter, George Bromberg, uses the client name in the first two words of his headline. I always preach to both writers and would-be writers that you seldom can beat an *incongruous-situation illustration or photo* for top readership. And that's exactly what Inland Steel resorted to here.

Imagine running a logo upside down. This wasn't done for shock appeal or just plain effect. Rather it was done because, darn it all, you can't build a bird's nest on *top* of a hard hat—you've got to invert it. Any smart bird

facilities—all are nearing completion. What's more, the current program includes an up-to-the-minute quality control laboratory. Throughout its more than 70 years of steelmaking, Inland has made its headquarters in Chicagoland, shared in the phenomenal growth of the midwest, anticipated its needs. Inland is here, looking toward the future . . . with great expectations.

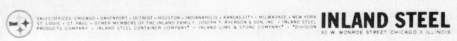

will peep you that! Inland's manager of marketing communications, Richard Killelea, worked with Bromberg in producing this series of 12 four-color spreads in order to fend off the encroachement of competition in the Midwest, which is Inland Steel's backyard. The entire campaign ran about 20 years ago and it obviously worked because Inland is still doing a big business in Chicagoland. At the time it ran, Inland's readership scores hit the stratosphere. The company's sales reps loved it, and tracking studies assured the company a dominant place in the annals of the business-to-business advertising fraternity. This was the first ad in the series and, quite logically, the very next one showed a beautiful white pigeon on *top* of the hard hat—so the logotype was finally righted.

International Harvester

Fletcher/Mayo Associates, the agency for International Harvester's Construction Equipment group, produced this excellent four-color spread that was part of a campaign that won them an American Business Press O&R Award. Let's face it, you just can't beat the pulling power of a fine photograph, especially when it's in full color. You can bet there's nothing phony about the behemoth International Harvester skidder grappling with a tall tree in this eye-catching ad. Every tyro copywriter and art director should know that nothing attracts business/professional buyers

more than a dramatic, action-packed shot of a product being used right on the job.

We all learn in grade school that use of action words in composition writing always brings in better marks from English teachers. The copy in "Skid more timber to the landing" is packed with action words. That's what makes it such good reading. And note that the copy also uses the jargon or patois of the timberman. To get good readership, it always pays to use the vernacular of the audience you're trying to reach. International Harvester picked the Forestry market as a primary market the year this ad ran. As a result, the company's S8A skidder increased its market shared by 10% *in a market down 35%*. The S10 skidder increased its market share 5% *in a market down 27%*.

International Paper

According to *Adweek,* which recently honored this fine four-color spread, the little piggy in this ad for boxes by International Paper drew so much fan mail he avoided his trip to the abattoir. Apparently International Paper purchased "piggy" from a farmer for $35, and he now lives in semiretirement in New Jersey.

The ad is one of a continuing series of product-oriented corporate ads Ogilvy & Mather's copy and art team (Bill Fuess and Richard Hovenack) have been producing for IP. Philip Farin, manager of advertising and marketing publicity, claims that the ad brought lots of new business—one lead generated by one insertion in *Business Week* resulted in a new customer who spent more money than the entire produc-

Now this little piggy can go to market fresher than ever in a special moisture-resistant container by International Paper

How do you make sure fresh pork gets to market fresh, short of having to ship live pigs?

International Paper has found the ideal solution—the specially designed, moisture-resistant container that's floating our little piggy across the page.

Why does it have to resist moisture? Because fresh pork contains more moisture than beef and most other fresh-cut meats. That makes fresh pork a lot more difficult to ship.

During shipping, moisture "weeps" out of the pork and into the container.

What's more, when all that moisture hits the cold air inside refrigerated trucks and railcars, humidity really builds up. The conventional fresh pork container will soak up that humidity as well as the "weeping" juice to become soggy and weak.

A crushing problem

Yet, meat containers need to be strong, because each container is packed with from 60 to 80 pounds of fresh pork. Then they're usually stacked six or seven high for shipment. In the three days it usually takes to get from processing plant to supermarket, those soggy containers, especially the ones at the bottom, can crush and break open.

Solved by Weathertex

International Paper saw the problem and found the solution—right on our own shelves. It was a container made with Weathertex board, a tough, moisture-resistant container board we'd already specially developed to ship fresh chickens packed in ice.

Tests showed that this same material was perfect for shipping moist wholesome fresh pork, too.

Now meat-packers using IP containers have better assurance their fresh pork will always get to market in the freshest possible condition.

If you have a shipping or packing problem, let us take a fresh look at it. Our industry specialists may have the perfect solution ready and waiting, like Weathertex board for fresh pork. Or we can engineer a solution especially for you. It may be a special coating, or medium, or even a complete packaging system to reduce costs and handling.

A container to keep fresh pork fresher is only one example of International Paper's creative solutions to our customers' problems in every industry—from meat-packing to building to energy to packaged goods to the medical field and more. May we help *you?*

INTERNATIONAL PAPER COMPANY
220 EAST 42ND STREET, NEW YORK, NEW YORK 10017

Special International Paper containers keep their strength despite moisture and humidity. Ordinary containers can become soggy and crush.

We don't expect meat-packers to float pork to market. We've launched this piglet to point out that our IP container will resist moisture during shipping of fresh pork and get it to market in the peak of freshness.

---

tion and media cost of the ad. And, in subsequent years, this customer's business paid for the entire ad series. And well it should. Have you ever seen a more doleful but beautiful photograph? What a wonderful way to get attention! Just float a pig in your product and watch readership soar!

You not only get stopping appeal, you get an instant demonstration of the ruggedness of your container. No wonder the ad has constantly scored in the top 10% in broad business magazines wherever it has been studied. And no wonder, too, that in the packaging magazines it was a top scorer almost every time it ran. It even scored highest in the book in one magazine on its tenth insertion. Now that's being piggy!

Till death or drinking do them part.

Love used to conquer all. But today, roughly one out of every dozen marriages in America comes apart over drinking.

Why should this be any concern of business?

Well, apart from the fact that these statistics are all <u>people</u>, alcoholism is proving expensive.

The latest estimate is that alcoholism costs U.S. business some $19 billion a year.

Eight years ago at ITT, we created a program to try and help employees or their families at participating ITT companies who had alcohol-related problems.

Or who had other personal crises, anything from drugs to finances.

One conspicuous feature of the ITT Employee Assistance Program is a telephone hotline, which employees or their families can call 24 hours a day, 7 days a week.

Our counsellors listen and refer the caller to someone nearby who can help.

Since it began, this ITT program has come to the aid of several thousand of our people.

Most of them, we're glad to report, have been helped back to productive, even happier lives.

And equally to the point, most are still with us.

The best ideas are the ideas that help people. ITT

For more information, write to Director-ITT Employee Assistance Program, Personnel Department, 320 Park Avenue, New York, N.Y. 10022.

©1981 International Telephone and Telegraph Corporation

ITT

About 15 years ago, ITT, a vast, diversified, multinational, corporate giant, decided it wanted to unify the advertising done by all its divisions and subsidiaries. I understand ITT researched over 17,000 individual advertisements in its quest for a new advertising format. Once the format was decided on, all of ITT's agencies were instructed to adhere to it.

The dictates were simple—a rule-lined border, large photograph at top, uppercase and lowercase headline using Palatino extra bold type, followed by 12-point Stymie light body text and the ITT logotype in the lower right-hand corner. This tried-and-true format has been used successfully by ITT ever since. And, as shown, when the art director (Howard Benson) chooses a good, clean photograph and the copywriter creates a fine headline and body text to go along with it, you have a powerful advertisement.

Incidentally, the Needham, Harper & Steers copywriter, Barry Biederman, obviously has studied "The Art of Readable Writing" by Rudolf Flesch. Using the "Flesch formula" of cutting down sentence length, shortening paragraphs and average word length, and utilizing a large number of personal words and sentences, he makes the copy truly engaging. "Till death or drinking do them part" ended up in presidential wife Nancy Reagan's lap when a HEW officer used it to show the White House what the private sector was doing to help fight alcoholism. ITT made a lot of points and got a bunch of accolades from the executive branch when the National Council of Alcohol Abuse insisted on running this ad over its own NCAA logo. It shouldn't hurt.

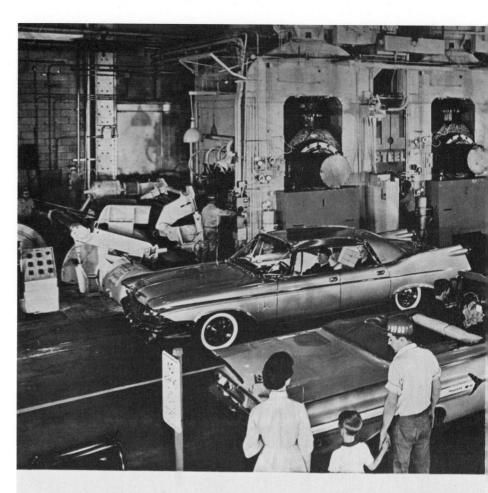

Travel, Adventure, Excitement— 76¢/lb.

The auto industry now delivers its end product at about 76 cents per pound—a bigger and better bargain than ever before. This bargain has been made possible by continuous improvements in design, in production techniques and in materials.

The auto industry has demanded more steel and more specialized steel—cold rolled steel in wide continuous strips, new alloy steels and new methods of heat treatment. The typical car today requires over 100 different grades of steel, and J&L now has the capacity to produce for the auto industry 68 per cent more steel than 10 years ago. The cold reducing mill

J&L Steel

This ad appeared in April 1962 in *Business Week*, *Time* and the *Wall Street Journal*. That's a lot of circulation, but more than 99% of it was sheer waste. Most of it was wasted because the ad was only meant to be read by the management people at General Motors, Ford, Chrysler and American Motors. The idea was to motivate the top hierarchy at these big steel-using automobile companies to buy Jones & Laughlin sheet steel. At that time, U.S. Steel and Bethlehem had a near-stranglehold on the market.

Jack Hight, the director of advertising of J&L, and Ray Johnson, the creative honcho at T. N. Palmer & Company, J&L's agency, came up with a campaign strategy that was directed to the steel company's customers' customers. The sales proposition was this: show consumers just how steel fits into their daily lives and why they should look for products made of steel when they buy. J&L felt that if it spent a

Traffic scene staged at new four-stand tandem cold reducing mill in J&L's Cleveland Works.   d'Arazien

behind the "traffic scene" rolls out 3,750 feet of steel sheet per minute — smooth, flat sheets of uniform thickness, held uniform by automatic X-ray.

In the last 10 years, Jones & Laughlin has invested nearly 700 million dollars in new plants and equipment to give *all* industry the improved steel needed to produce a better bargain. J&L, today, is a leading source for a great number of standard and specialty steels—sheets for the auto, appliance and electronic industries, stainless steels for the chemical industry, tubular products for the oil and gas industry and countless other specialized steels for countless other specialized industries.

Nearly every product made of steel is a better value today than it was just 10 years ago. These products get a better start with the strength, toughness and precision quality of Jones & Laughlin steel.

This Steelmark identifies the better value of products made of steel. Place this mark on your steel products—and look for it when you buy.

Jones & Laughlin Steel Corporation
PITTSBURGH, PENNSYLVANIA

JL STEEL

This advertisement is scheduled to appear in Business Week — August 27, Automotive News — August 22, Automotive Industry — August 15, Detroit Athletic Club News — August, 1960.

lot of advertising dollars to help sell cars, housewares and hardware, it would really win big brownie points with the manufacturers of these products who were big steel buyers. "Travel, Adventure, Excitement—76¢/lb" really packed a wallop. Art d'Arazien, the doyen of steel photographers, lit up and staged this entire traffic jam in J&L's rolling mill in Cleveland. A tough program logistically, but it paid off at the readership bank. The ad Starched a Noted of 69, Seen Associated of 64 and Read Most of 19 (the highest readership of any advertisement) the first time it ran in *Business Week*. Also, the agency was smart enough to put the arm on the *Wall Street Journal* and it became the first ad ever to run in the newspaper as a gutter-bleed spread. The results were spectacular. Bunkie Knudsen, General Motors chairman, lauded it as "the best damn car ad I've ever seen—wish we'd run it ourselves." Of the four big automobile prospects, two were sold. General Motors and Ford gave J&L directly traceable orders. The stranglehold was broken, for a time at least.

Solar power paint, from Johnson Wax polymers?

Someday, anything you paint—your home, for example—will be able to generate its own electricity directly from the sun's rays. Solar power paint, made with tiny photovoltic cells suspended in our special electricity-conducting liquid polymer, would make today's energy crisis just a distant memory. It's still a "someday" idea. But it may be some day soon.

Before long, other special polymers, in conjunction with other materials, might make matches wind and weatherproof, or make cans bio-degradable.

Today, special Johnson Wax polymers work in such widely diverse applications as coating matchbooks, improving printing ink, and helping wall panels radiate heat.

Such polymers can make practically anything you make—of metal, paper, plastic, fabric, wood or leather—look better, stand stronger, last longer or weather less.

What can one of our special polymers help you do, right now? Only you can tell us. Then you'll see why our business is innovative chemicals for professional use. We are the Specialty Chemical Group, Worldwide Innochem Operations, Johnson Wax, Racine, WI 53403. We love a challenge. Call or write us with yours.

Johnson wax
SPECIALTY CHEMICALS

We do much more than floors.

Johnson Wax

The specialty chemicals people at Johnson Wax employ the hubristic copy approach (in sort of a boast-and-claim style) to educate the uninformed that "we do much more than floors." The company name has been a household word for years but, unfortunately, nobody in industry recognizes Johnson Wax as a factor in the specialty chemicals business. David Klaproth, Campbell-Mithun's copywriter, and his compatriot art director, Bob Ritz, came up with this futuristic awareness campaign to change that forthwith.

The best in this series of four-color advertisements is this electric-light-bulb-in-a-can-of-paint single-pager. The first paragraph very quickly informs the reader what Johnson's special polymers will be able to do "someday." (In this case, house paint containing the company's polymer will contain electricity-conducting photovoltaic cells which will generate electricity from the sun's rays). Someday. That means the Johnson Wax people are heavy into research and development. The copy then goes very logically from "Someday" to "Today." The third paragraph spells out the great things presently being done by Johnson's specialty chemicals. The trick photography is exquisitely done, and it's unusualness makes the reader stop and stare—then go on to read the copy.

DROODLES
by ROGER PRICE

"RECORD BUG GETTING INTO THE GROOVE"

This provocative Droodle is one of my finest works. One of the toughest jobs I've ever tackled, too, because I'm a stickler for the authentic, and stood poised over the spinning disc for four hours, in order to capture the bit of action so beautifully depicted above.

Our subject is a flea named Sam. Quite a character, but *talented!* Sam is the only flea ever to study music at the Juilliard Academy. Had to leave in his junior year, though — got into a jam with an accordion teacher.

Anyway, Sam listened and listened to this record — "Doggie in the Window" — and then sprang in to investigate. When he discovered that the doggie under discussion was a Mexican Hairless, he took a powder and hasn't been seen since. Just as well, though, for he was due for a brush-off.

Don't *you* get a brush-off! Get in the groove and help your production department set new records for efficiency! How? Investigate how the Jones & Lamson Optical Comparator can speed up and improve inspection operations in your plant. J&L Comparators are used throughout industry to measure and inspect all sorts of objects and parts ranging in size from the tiniest screw to large turbine blades.

They make inspection and measurement easy and rapid, and they're accurate to .0001".

Learn more — send this coupon today!

Punched holes in radio tube micas being inspected on a Jones & Lamson Optical Comparator. J&L makes a complete line of 11 Comparators, in both bench and pedestal types

JONES & LAMSON

JONES & LAMSON MACHINE COMPANY, Dept. 710, 529 Clinton St., Springfield, Vt., U. S. A.

☐ Please send me your booklet, "Comparators — what they are and what they do".
☐ I'd like to see a showing of the movie, "What's the Difference",

on _____ (date)
Name _____ Title _____
Company _____
Street _____
City _____ Zone ____ State _____

Jones & Lamson

Sometimes just by being different you can come up with a campaign that gets results. Copy chief Leonard Linnehan of Boston's Henry A. Loudon agency came up with this fine campaign for Jones & Lamson. The company wanted to find new markets for its optical comparator, a machine used for inspecting and measuring all kinds of industrial things. The client, Norman Richardson, manager of marketing services, decided to run a series of ads in *Scientific American* to pick the brains of long-haired engineering types. Linnehan got the bright idea that a Roger Price "Droodles" series would appeal to this very sophisticated audience. In the mid 1950s Roger Price was somewhat of a national institution. His humorous "Droodles" drawings and offbeat philosophical commentaries were the bases for a highly rated television show. "Droodles" was also a very popular nationally syndicated newspaper column.

    The campaign was an instant success insofar as readership, response and subsequent sales were concerned. Jones & Lamson was inundated with coupon returns. Sixty percent of these were accompanied by letters of thanks from engineers saying they loved the ads because at long last somebody thought of them as real people, not robots.

    The campaign was expanded to run in a bunch of trade publications like *Industrial Equipment News, New Equipment Digest, American Machinist, Tooling and Production* and many others. After running two or three years, brand preference studies showed J&L to be number one by a whopping 58%. Sales from offbeat new markets came in like gangbusters. Machines were sold to the New York District Attorney's office. The Fish and Game Division of the Department of Interior bought one to inspect and measure the scales of fish. And to top it all off, a textile outfit purchased one to measure the contact closeness of a line of brassieres.

How to solve your rock-hauling problems at low cost

Large bowl target (15'4" x 10'2" on B Rear-Dump) permits easy loading without spillage. Open rear of body provides wide, low entry for dipper — to give extra speed advantage to your shovel.

With today's highway specs calling for *straight, level* right-of-ways, earthmoving contractors are often faced with the problem of removing large rock formations. With LeTourneau-Westinghouse Tournapull® Rear-Dumps, you can maintain job efficiency and still keep equipment inventory at a minimum.

These high-production L-W haulers offer plenty of power, traction and maneuverability for toughest off-road earthmoving. What's more, you have the economy advantage of being able to interchange scraper, flat-bed, crane, etc., for Rear-Dump unit — behind same prime-mover.

Check the following Tournapull Rear-Dump features. They can help solve your rock-hauling problems ...and increase profits!

Hauls anywhere — Big single, *load-rated* tires let Rear-Dumps easily haul cross-country — over roughest terrain, through muck and soft fills. Unit also travels job-to-job under its own power, over paved highways or city streets. Tires do not damage paving, RR tracks, etc.

Speeds loading — Wide bowl, with low rear entry, makes loading easier, faster. Permits quick bucket swing-out, while dipper-door is still open. And because of large target area, spillage is greatly reduced.

← Front-wheel drive Tournapull hauler dumps load clean over bank edge, eliminating dozer clean-up. Pulls out with front wheels on solid footing. Can also "hump" out of soft footing without wheel spin — by dump action plus alternate braking of front and rear wheels.

Production delay for clean-up around shovel is minimized.

Dumps fast, clean — Just the flick of dashboard switch instantly activates point-of-action electric hoist motor. Body raises quickly to desired angle. At full dump position, edge of bowl is low behind rear wheels ... so material cannot roll forward to lodge against wheels, nor pile under rear end. Smooth body sheds stickiest material readily.

Resists body shock, damage — Slanting walls of all-steel body, plus heavy 3-layer bottom, beat shock-load problems. Heaviest materials deflect off sloping sides ... quickly build up a shock-cushion in bottom. Reinforced bowl floor on C Rear-Dump — for example — consists of $1\frac{1}{2}$" steel billets, between $\frac{1}{2}$" steel bottom and a $\frac{3}{4}$" steel plate facing.

Short 180° turns — Rear-Dump makes continuous 180° turn in space less than hauler's own length. In dump position, it turns in only about $\frac{2}{3}$ of overall length! This unusual maneuverability of L-W Rear-Dumps allows you to work in tight quarters where smaller conventional haulers often cannot go.

For complete information

Why not see how you, too, can increase production, lower hauling costs with L-W Tournapull Rear-Dumps? Call or write for complete information on price, delivery, and the bonus interchangeability features. 3 sizes: 11, 22 or 35 tons.

R-1448-DC-1

 LeTOURNEAU-WESTINGHOUSE COMPANY, PEORIA, ILLINOIS

Subsidiary of Westinghouse Air Brake Company

Where quality is a habit

Le Tourneau-Westinghouse

Way back in the early 1950s, Le Tourneau-Westinghouse was manufacturing and selling a wide variety of earthmoving machines for the mining and construction industries. The line consisted of 23 different classes of equipment, all big in size as well as price. The advertiser was running two-color spreads showing just a few pieces of equipment and telling what the product was rather than what it did. Obviously, with so many major items demanding promotion, ads devoted to one or two products under a single job condition were not doing the job. No one product was getting sufficient attention over a year's time.

Joe Serkowich, Le Tourneau's advertising manager, and his agency's copy chief, Larry Roth, came up with a four-page pictorial campaign with striking results. The new ads featured as many as seven pieces of equipment in two black and white, consecutive double-page spreads. In the first ads that ran, eleven different jobs were illustrated, five nations were represented and six markets covered. Rather than confuse the readers of magazines like *Engineering News Record* and *Construction Methods and Equipment,* the ad scored a whopping reader rating of 72% (the old ads were in the 20% to 30% range). Overall, the new campaign jumped reader interest two-and-a-half times. Serkowich has been credited by many of his peers as the founder of the *multiple-page picture-ad* technique which is not dissimilar to formats used in magazine editorials. Today this is called *advertorial style.* Note how the layout is strictly editorial; orderly arrangement predominates, and there is very little or no "addy" feeling at all. The logo is absolutely hidden, and the use of one dominant photograph really draws your attention to the others which show what the product does on the job site. Also, the photographs are of scenes that the reader is really interested in.

Over the last 27 years this fine campaign and its format have been copied to a fare-thee-well. A well deserved honor, indeed.

This is 80% of Railroading

It is sometimes well to back off
and take a look at the overall picture.

We're thinking of the steam locomotive. These locomotives
are producing three *billion* ton-miles—and will do it again tomorrow
and the next day. They—these steam locomotives—are doing 80 per
cent of the work on the railroads—more work than
they ever did in any year before 1941.

Many of these locomotives are old, too old, and have distorted
the statistics on performance. Many, however, are modern. And
on modern steam power—locomotives that pack 5000 to 9000
horsepower and can stay on the road for 16 and 18 hours, and then
turn around in an hour or two—the statistics look pretty good.

We build such modern power—and are convinced that it has its place.

DIVISIONS: *Lima, Ohio*—Lima Locomotive Works Division; Lima Shovel and Crane Division. *Hamilton, Ohio*—Hooven, Owens, Rentschler Co.; Niles Tool Works Co. *Middletown, Ohio*—The United Welding Co.

PRINCIPAL PRODUCTS: Locomotives; Cranes and shovels; Niles heavy machine tools; Hamilton diesel and steam engines; Hamilton heavy metal stamping presses; Hamilton-Kruse automatic can-making machinery; Special heavy machinery; Heavy iron castings; Weldments.

April 30, 1949

Lima Hamilton

A while back a reporter at a B/PAA convention asked Cahners Publishing's research VP John DeWolf what he considered the most important development in business-to-business advertising. John reflected on his almost 50 years of association in the business and claimed it was the original McGraw-Hill repeat ad study, and the subsequent one done by the Industrial Advertising Research Institute. He may well be right. Until the first study came along, most advertisers never repeated advertisements. Even today, a plethora of advertisers are reticent to repeat—even once. DeWolf made his reputation as a readership expert while at the old G.M. Basford ad agency where he wrote ads like this great one for Lima Hamilton. In those days he wrote a new Lima ad every week because the top book of the industry, *Railway Age*, was a weekly publication.

This particular one appeared in 1949 and to the railroad buff it was, and still is, a classic. At first glance you might say the headline is somewhat anemic. But to the railroader it got to the jugular fast by pointing out that 80% of the work on railroads was being done by steam locomotives. The headline was quite artistically coupled to the beautiful photograph of the steam locomotive. The copy reads like a sonnet: "It is sometimes well to back off and take a look at the overall picture.... Many of these locomotives are old, too old, and have distorted the statistics on performance. Many, however, are modern. And on modern steam power—locomotives that pack 5000 to 9000 horsepower and can stay on the road for 16 and 18 hours, and then turn around in an hour or two—the statistics look pretty good."

Unfortunately DeWolf's ads didn't save these denizens of the horse-and-buggy era, but his copy was pure magic to the ears of the railroad aficionados. I'm not so sure the Long Island Railroad and others wouldn't have better on-time records today if they still had a few of these locomotives pulling trains.

Loctite Pipe Sealant with Teflon stops leaks for good.

Even where tape fails.

- Loctite PST won't let fittings vibrate loose like tape—yet can be taken apart with a pipe wrench.
- Unlike tape, Loctite PST completely fills the "leak-paths" in threads to make a leak-proof seal—also prevents thread corrosion.
- Unlike tape, Loctite PST never shreds to contaminate system.
- Loctite PST seals instantly.
- Loctite PST is easy to apply.
- Loctite PST costs less per application than tape.

Free sample and Leak Cost Calculator available—write Loctite Corporation, Newington, Connecticut 06111.

"Stop Leaks" seminar available through your distributor. For his name and phone number, call toll-free 1-800-243-8810 (CT 1-800-842-8684).

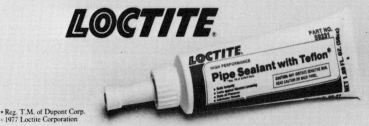

*Reg. T.M. of Dupont Corp.
© 1977 Loctite Corporation

Loctite

Who says comparative advertising doesn't work? Mintz & Hoke, a smallish New England agency, got its start in industrial advertising by producing super ads like this one for Loctite Corporation. The headline and this very dramatic photograph instantly show the difference between ordinary pipe tape and Loctite pipe sealant with Teflon. The difference of course is that the joint sealed with pipe tape leaks, and the one sealed with Loctite doesn't.

"Loctite Pipe Sealant with Teflon" ran in four colors in *Industrial Equipment News* and *Plant Engineering* in 1977 and generated over 7,000 sales leads in four months. Copywriter Dick Haddad and Art Director Rick Whittey should be proud to note that this single effort ran up the highest readership score in six years in *Industrial Equipment News*.

Mintz & Hoke was smart enough to reprint a bunch of these ads in an agency mailer and send it out to new business prospects. At the time, Loctite was M & H's sole industrial account. Today, the agency has seven industrial clients representing 35% of its $16 million annual capitalized billings. It all goes to prove that if you produce great ads you can't help but succeed.

They Flew a Brave Flag

...and we must recreate that spirit today.

They brought with them brave new ideas, the men and women who started America... Freedom... Self-reliance... Independence.

And when the Old World threatened their freedom, they flew a brave new flag that warned, "Don't Tread on Me."

They backed up their brave flag by working together and sharing the load. They respected and trusted each other, and worked together for a common cause.

They were all workers, with no free loaders. All were expected to carry their share of the burden.

We need to fly that Colonial flag again today. We need to declare our economic independence, and rekindle the courage and resourcefulness that built America.

Let's march ahead together under a brave flag... let's get together, and work hard together to rebuild our nation.

With new resolve, we must succeed.

James E. Stewart

James E. Stewart, Chairman of the Board and Chief Executive Officer
Lone Star Industries, Inc., One Greenwich Plaza, Greenwich, Connecticut 06830

LONESTAR

Number One in Cement... Serving America's Great Builders.

Would you like a "Don't Tread on Me" flag?

Miniature desk flags—limited quantity available printed in color on rayon, 4" x 6" in size. Send check for $1.00 for each flag, including postage and handling costs. Mail coupon and remittance to: Lone Star Industries, Inc., c/o Accurate Mailing Service, 215 Route 10, Dover, N.J. 07801.

Name_____

Address_____

City_____ State_____ Zip_____

Quantity desired_____ at $1.00 each. Amount enclosed_____

Lonestar

Lonestar went through a number of pretty damn good ad agencies before coming up with this fine *Wall Street Journal* corporate campaign. The series was conceived and produced in-house by Ted Price, Lonestar's director of advertising and creative services. Oddly enough, the ads themselves break a lot of rules. For one thing, they're very busy. This particular effort contains a number of disconcerting elements that seem to be just thrown together: two flag illustrations (with type on each of them), a lot of different type weights and type sizes, a handwritten signature, a big logotype, a "designy" symbol, a slogan—and to bottom it off, a coupon! But darned if the whole format doesn't make you *want to read* what the advertisement is all about. I guess it's just *so* busy and *so* old-fashioned that it draws your complete attention directly to the page like a magnet.

We all know in the early eighties that nostalgia, patriotism and Reaganesque-conservatism is terribly *in*. And that's what makes this advertisement work. When Lonestar first ran the ad, Ted Price anticipated requests for a thousand or so flags. And at a buck a clip you'd guess he was being conservative. After just one insertion in the *Wall Street Journal,* Lonestar has been deluged with dollars. In fact, they have received requests for 15,000 flags and they're still coming in. Let's face it, you just can't argue with success.

Lukens Steel

The specialty steel people from Coatesville, Pennsylvania have a right to be proud about this 1977 four-color ad spread that was prepared in Marsteller's New York office by Bob Singer with art direction by Rich Redmond. As in great ad after great ad, the picture tells the whole story and, in doing so, makes this an outstanding ad. One of the reasons the photograph stops you and makes you want to read further is because you immediately relate to that pretty little inquisitive-looking girl. Anyone who has ever sipped a soft drink through a straw has also, at one time or another, blown into the straw to make bubbles. So when you see a picture of someone else doing it at perhaps the same age you did

Introducing Fineline.™

Lukens is using a basic law of physics to improve the quality of plate steel.

By blowing calcium particles through molten steel, much like blowing bubbles with a straw, Lukens is now able to produce large volumes of low-sulfur plate steels.

It's a simple but revolutionary steelmaking process we call Fineline.

Through this process, the sulfides and oxides normally present in a steel melt are greatly reduced. Those that remain have a uniform roundness, called "shape control," and are more uniformly distributed. The result is improved ductility in all three directions of a rolled steel plate, especially the through-gage or Z-direction.

The Fineline process also develops improved notch toughness. So designers can now set tighter specifications necessary for critical plate steel applications.

There's a lot more to be said about Fineline steels. You'll find it in our new brochure. Write or call Bob Sterne, Market Development Division, Lukens Steel Company, Coatesville, Pa. 19320 (215) 383-2404.

Lukens Steel
The specialist in plate steels

---

it, you immediately stop, look and read. You've been there yourself.

Lukens wanted to introduce its new Fineline process which works by blowing calcium particles through molten steel with a very low sulfur content. This super ad, photograph and crisp copy, does the job beautifully. David Hungerford, Lukens long-time advertising and product information manager, deserves a special accolade because the ad launched a product line that has contributed heavily to Lukens' total sales. The public relations experts liked the photograph so much they used it on Lukens' 1977 Annual Report. Not only that, but the ad was named the top four-color spread in the Business/Professional Advertising Association's prestigious 1977 Pro-Comm Awards competition.

Human beings come in all sizes, a variety of colors, in different ages, and with unique, complex and changing personalities.

So do words.

There are tall, skinny words and short, fat ones, and strong ones and weak ones, and boy words and girl words and so on.

For instance, title, lattice, latitude, lily, tattle, Illinois and intellect are all lean and lanky. While these words get their height partly out of "t's" and "l's" and "i's", other words are tall and skinny without a lot of ascenders and descenders. Take, for example, Abraham, peninsula and ellipsis, all tall.

Here are some nice short-fat words: hog, yogurt, bomb, pot, bonbon, acne, plump, sop and slobber.

Sometimes a word gets its size from what it means but sometimes it's just how the word sounds. Acne is a short-fat word even though pimple, with which it is associated, is a puny word.

Puny words are not the same as feminine words. Feminine words are such as tissue, slipper, cute, squeamish, peek, flutter, gauze and cumulus. Masculine words are like bourbon, rupture, oak, cartel, steak and socks. Words can mean the same thing and be of the opposite sex. Naked is masculine, but nude is feminine.

Sex isn't always a clear-cut, yes-or-no thing on upper Madison Avenue or Division Street, and there are words like that, too. On a fencing team, for instance, a man may compete with a sabre and that is definitely a masculine word. Because it is also a sword of sorts, an épée is also a boy word, but you know how it is with épées.

Just as feminine words are not necessarily puny words, masculine words are not necessarily muscular. Muscular words are thrust, earth, girder, ingot, cask, Leo, ale, bulldozer, sledge and thug. Fullback is very muscular; quarterback is masculine but not especially muscular.

Words have colors, too.

Red: fire, passion, explode, smash, murder, rape, lightning, attack.

Green: moss, brook, cool, comfort, meander, solitude, hammock.

Black: glower, agitate, funeral, dictator, anarchy, thunder, tomb, somber, cloak.

Beige: unctuous, abstruse, surrender, clerk, conform, observe, float.

San Francisco is a red city, Cleveland is beige, Asheville is green and Buffalo is black.

Shout is red, persuade is green, rave is black and listen is beige.

Oklahoma is brown, Florida is yellow, Virginia is light blue and Massachusetts is dark green, almost black. Although they were all Red, at one point Khrushchev was red-red, Castro orange, Mao Tse-tung gray and Kadar black as hate.

One of the more useful characteristics of words is their age.

There's youth in go, pancake, hamburger, bat, ball, frog, air, surprise, morning and tickle. Middle age brings abrupt, moderate, agree, shade, stroll and uncertain. Fragile, lavender, astringent, acerbic, fern, velvet, lace, worn and Packard are old. There never was a young Packard, not even the touring car.

Mostly, religion is old. Prayer, vespers, choir, Joshua, Judges, Ruth and cathedral are all old. Once, temple was older than cathedral and it still is in some parts of the world, but in the United States, temple is now fairly young. Rocker is younger than it used to be, too.

Saturday, the seventh day of the week, is young while Sunday, the first day of the week, is old. Night is old, and so, although more old people die in the hours of the morning just before the dawn, we call that part of the morning, incorrectly, night.

Some words are worried and some radiate disgusting self-confidence. Pill, ulcer, twitch, itch, stomach and peek are all worried words. Confident, smug words are like proud, lavish, major, divine, stare, dare, ignore, demand. Suburb used to be a smug word and still is in some parts of the country, but not so much around New York anymore. Brooklyn, by the way, is a confident word and everyone knows the Bronx is a worried word. Joe is confident; Horace is worried.

Now about shapes.

For round products, round companies or round ideas use dot, bob, melon, loquacious, hock, bubble and bald. Square words are, for instance, box, cramp, sunk, block and even ankle. Ohio is round but Iowa, a similar word, is square but not as square as Nebraska. Boston is, too—not as square as Nebraska, but about like Iowa. The roundest city is, of course, Oslo.

Some words are clearly oblong. Obscure is oblong (it is also beige) and so are platter and meditation (which is also middle-aged). Lavish, which as we saw is self-confident, is also oblong. The most oblong lake is Ontario, even more than Michigan, which is also surprisingly muscular for an oblong, though not nearly as strong as Huron, which is more stocky. Lake Pontchartrain is almost a straight line. Lake Como is round and very short and fat. Lake Erie is worried.

Some words are shaped like Rorschach ink blots. Like drool, plot, mediocre, involvement, liquid, amoeba and phlegm.

At first blush (which is young), fast words seem to come from a common stem (which is puny). For example, dash, flash, bash and brash are all fast words. However, ash, hash and gnash are all slow. Flush is changing. It used to be slow, somewhat like sluice, but it is getting faster. Both are wet words, as is Flushing, which is really quite dry compared to New Canaan, which sounds drier but is much wetter. Wilkinsburg, as you would expect, is dry, square, old and light gray. But back to motion.

Raid, rocket, piccolo, hound, bee and rob are fast words. Guard, drizzle, lard, cow, sloth, muck and damp are slow words. Fast words are often young and slow words old, but not always. Hamburger is young but slow, especially when uncooked. Astringent is old but fast. Black is old, and yellow—nearly opposite on the spectrum—is young, but orange and brown are nearly next to each other and orange is just as young as yellow while brown is only middle-aged. Further, purple, though darker than lavender, is not as old; however, it is much slower than violet, which is extremely fast.

Because it's darker, purple is often softer than lavender, even though it is younger. Lavender is actually a rather hard word. Not as hard as rock, edge, point, corner, jaw, trooper, frigid or trumpet, but hard nevertheless. Lamb, lip, thud, sofa, fuzz, stuff, froth and madam are soft. Although they are the same thing, timpani are harder than kettle drums, partly because drum is a soft word (it is also fat and slow) and as pots and pans go, kettle is one of the softer.

There is a point to all of this.

Ours is a business of imagination. We are employed to make corporations personable, to make useful products desirable, to clarify ideas, to create friendships in the mass for our employers.

We have great power to do these things. We have power through art and photography and graphics and typography and all the visual elements that are part of the finished advertisement or the published publicity release.

And these are great powers. Often it is true that one picture is worth ten thousand words.

But not necessarily worth one word.

The *right* word.

The Wonderful World of Words

MARSTELLER INC.

Burson-Marsteller · Marsteller International · Medical Group · The Zlowe Company   Advertising · Public Relations · Marketing Research   Chicago · Los Angeles · New York · Pittsburgh · Washington, D.C.   Brussels · Geneva · London · Paris · Stockholm · Stuttgart · Toronto

Marsteller

A couple of years ago, Bill Marsteller sold his $300 million ad agency to Young & Rubicam for close to 30 million bucks. Since then, he's semiretired and taking life easy. He sure deserves it, but we all miss him—not only because he's one of the brightest ad men in the business, but also because we miss reading some of the wonderful copy he consistently wrote (his book of interoffice staff memos is a classic). His "greatest" ad is this house ad which ran in 1971 just once in the *Wall Street Journal.* To a writer, "The Wonderful World of Words" only creates envy. I must have read it 20 or so times, and each time I get sick with jealousy. Why didn't I think of it and write it? The copy describes the size and shape and color of words. It also describes the sex and physical characteristics and age of words. It actually humanizes words and in so doing forces you to read every single word of this fabulous advertisement about words. Did it work? Never fear. The same day the ad appeared, Marsteller's switchboard received a call from the president of Boyle-Midway's Division of American Home Products. He wanted to speak to the writer of the ad he had just read in the *Wall Street Journal.* Two weeks later Marsteller landed AHP as a $3 million account. Bill's art directors for this all-type ad were Al Herbrecht and Joe Goldberg.

"I don't know who you are.
I don't know your company.
I don't know your company's product.
I don't know what your company stands for.
I don't know your company's customers.
I don't know your company's record.
I don't know your company's reputation.
Now—what was it you wanted to sell me?"

MORAL: Sales start **before** your salesman calls—with business publication advertising.

McGRAW-HILL MAGAZINES
BUSINESS • PROFESSIONAL • TECHNICAL

McGraw-Hill

When I was a mailboy at Fuller & Smith & Ross, I used to deliver mail to Gil Morris, a vice president of that once-great ad agency. Gil was the model used in the first version of this famous "man in the chair" advertisement for McGraw-Hill. His perpetual scowl belied a rosy disposition and demeanor. Unfortunately, I was so far down the corporate ladder I was frightened to death every time I had to enter his office. And that frigid face (and others after it) really tells the story. The copy, "I don't know who you are," was exactly what McGraw-Hill has been getting across to the millions of readers of this ad since 1958. Written by Henry Slesar, one of the finest copywriters of our time, this "greatest" industrial ad has been run in Russian, German, Italian, French and, yes, even in Chinese. The people of McGraw-Hill tell me that hardly a day goes by without requests for reprints or for permission to reproduce it for use in a book or magazine article. The company prints about 15,000 copies of the ad a year just to handle the demand. McGraw-Hill's Publications Division makes a lot of money out of the advertising business, and one of the reasons they do is because they spend a lot of money on advertising McGraw-Hill—wisely.

Ah, wilderness...

You've probably heard the tales before. The heroes and heroines are various species of wildlife, all facing extermination by the villains, who are, of course, rapacious industrialists. A favorite setting is Alaska, and the story has it that the 800-mile-long Trans Alaska Pipeline is "endangering" Alaska's caribou.

Well, take a good look at the picture. Those aren't oilmen dressed up in caribou costumes—those are bona fide, card-carrying caribou, playing themselves in the very shadow of the pipeline itself.

In fact, scientists studying the animals in situ have found that the Central Arctic Herd, the one that roams the pipeline's locale, has actually <u>increased</u> in number since the pipeline was built. In 1969, well before construction began, there were about 5,000 of them. Today, the herd numbers more than 7,000.

We're not saying that the pipeline is directly responsible for the animals' proliferation. But it obviously hasn't dampened their amorous inclinations either, or their appetite for the thick grasses planted beneath the pipeline by the companies that built it. The caribou forage beneath the raised sections, as the photo shows.

So there's living proof that oil and animals can coexist. To contend otherwise would be to commit a caribouboo.

Mobil

Mobil pioneered the *opposite editorial* (op ed) newspaper format which had been copied by a number of business-to-business advertisers. The company is known as a tough media street-fighter that goes all out to explain oil industry profits, pricing and management decisions in its advocacy advertising campaign. The ads run in a number of the country's big circulation daily newspapers. Most of the ads that run are of the all-type variety, but from time to time Mobil changes this just to break the monotony. What gives the ads dominance is the two-column editorial-type format with the small Mobil logotype centered underneath. This, plus the op ed position and weekly continuity, achieves fantastic readership impact. My favorite is this one showing a herd of caribou frolicking under the trans-Alaska pipeline. The copy makes fun of Mobil's detractors who liken the company to "rapacious industrialists" whose main interest is to exterminate caribou. The glib text avers, "Well, take a good look at the picture. Those aren't oilmen dressed up in caribou costumes—those are bona fide, card-carrying caribou, playing themselves in the very shadow of the pipeline itself." When the reader gets through reading he has to go away with the feeling that big old Mobil can't really be half-bad, much less a profiteer! Public relations luminary Herbert Schmertz and Mobil's agency, Doyle Dane Bernbach, are responsible for the series.

WE USED TO BRING YOU THE SIX O'CLOCK NEWS. NOW WE ARE THE SIX O'CLOCK NEWS.

Back in the 1950s, when the 6 o'clock news first started finding its way into people's living rooms, Motorola was there.

Today, although we're no longer making television sets, we're still there when people all over the country sit down to the 6 o'clock news.

Except now, instead of making the sets they're watching, we're part of the history they're watching.

On November 12, 1980, a Motorola communications subsystem designed for the Jet Propulsion Laboratory and NASA sent photographs back to Earth from a billion miles away near the planet Saturn.

This equipment, the only link between Earth and the Voyager spacecraft, not only sent photos that thrilled a watching world, it also transmitted data that turned centuries of scientific thought upside down.

One month before, the world's largest auto maker was able to announce a giant step forward in improving gas mileage while at the same time decreasing emissions because of an engine management system that runs on a microprocessor Motorola designed.

And in the years ahead, technology we're pioneering may bring microelectronic devices into our lives that the world has only dreamed of before.

Like a portable telephone small enough to fit in your pocket.

Microprocessors that could allow industrial robots to function ten times faster than humans.

And microchips that make it possible to use computers to find oil miles beneath the earth's crust without drilling an inch.

All these advances and more are being pioneered by Motorola in our

Motorola

A beautiful play on words is used in the headline to make this corporate Motorola ad stand out. "We Used to Bring You The Six O'Clock News. Now We Are The Six O'Clock News" is all that's needed to tell the public that, while Motorola gave up making television sets 30 years ago, the company is still a very big factor in the electronics business. Young & Rubicam prepared this ad which shows a very warm gemütlich living room setting with a television set turned on to a picture of Saturn. The copy explains that a Motorola communications substation helped send the photograph back to Earth from a billion miles away in space. Signing off, the text makes it very clear to the

velopment centers around the world. And in many ways they're only the ginning of what we can do.

So the next time you see something on the 6 o'clock news that you never agined possible, you'll know there's a chance someone at Motorola had a part making it happen. And if that makes us sound like a company far different m the one that once made television sets for your living room, it's simply cause we are.

Making electronics history. MOTOROLA

reader that Motorola is far different from the company that once made television sets for the living room. The ad creator was very intelligent not to include people sitting around on the chairs and sofa watching the telly. All too often, art directors and photographers people ads to death just for the sake of avoiding boredom. What makes this ad great is that the photograph is uncluttered with pretty models who wouldn't have added an iota to its readership. People literally would have been a distraction to the main focus of the ad, which is the picture of Saturn on the boob tube. Great restraint! Created by art director Tom Shortlidge and copywriter Mike Fenske, the ad won a coveted Silver in the One Show in 1981.

At last.
East meets West.

Now, you can reach a détente between the subtle flavorings of Chinese foods and the realities of American canning.

All it takes is the combined experience, knowledge and ingredients of three experts in international relations: National and its subsidiaries, Seasonings Inc. and Scientific Flavors Inc.

With Col-Flo® 67, Hi-Flo,® Clearjel® and National® 465, you'll get the right mouth feel. Smooth, not stringy. Short, not chunky.

National Starch

At the risk of sounding duplicitous, I can truthfully say that, for as long as I can remember, National Starch advertisements have always Starched in the top ten percentile readership bracket. The combination of fine art direction (including excellent graphic design, superior photography and typography, along with some superb copywriting) really turns the trick.

"At last, East meets West" makes you salivate right on this four-color double-page spread. And what a gourmet spread it is. In

And your vegetables will stay suspended in the sauce, so your product will look appetizing too.

To enhance the subtle taste of Chinese food, we have our special waxy maize starches. They're perfectly bland, so the true flavor of the vegetables always comes through. And special formulations of our seasonings and flavorings will complement that taste.

We'll help you balance your sauces perfectly for manufacturing too. There'll be enough viscosity to avoid splashing during high speed filling, but not too much so your production is slowed down.

Don't let your ethnic foods create an international incident. Call or write National Starch & Chemical Corp., 10 Finderne Avenue, Bridgewater, New Jersey 08807. (201) 685-5000.

National STARCH AND CHEMICAL CORPORATION

Seasonings Inc. Scientific Flavors Inc.
The ingredients for success.

the first place, the informative, short, two-sentence headline runs right into the copy. There are no corny, cub copywriter tricks like hyphens, dots or dashes. That would be too tacky for Phil Costa, the Marsteller art director who laid out this one. Bob Singer, the agency creative director, wrote the headline and, with Costa's help, used it to point directly to the first copy block. This averts the problem of a reader or potential reader getting lost.

National Starch ads are not designed to lose readers. No way. The lifelike photography sees to that, and so does the scintillating copy. The short, easy-to-read, common-sense text reads like dialogue between two people. The writer and the reader. Smart.

☞ THIS IS PENN/BRITE OFFSET...THE VALUE SHEET

ABCDEFGHIJKLMNOP
QRSTUVWXYZ&**ABCD**
EFGHIJKLMNOPQR
STUVWXYZ&ABCDEF
GHIJKLMNOPQRST
UVWXYZ&ABCDEF
GHIJKLMNOPQRS
TUVWXYZ&ABCD
EFGHIJKLMNOP
QRSTUVWXYZ&

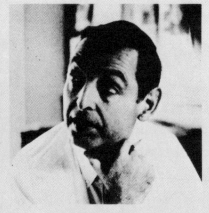

New York and Penn commissioned designer Herb Lubalin to create this insert, and to incorporate in it practically all the demands on the printability of paper which could be encountered. See for yourself how faithfully Penn/Brite Offset has reproduced each of his design elements ... how well it passed his "torture test." Penn/Brite Offset is the white, bright, *value* sheet that comes to you moisturized and double-wrapped. Write for our new, complimentary swatch book and the name of your nearest distributor to: New York & Pennsylvania Co., 425 Park Avenue, New York 22, New York

New York and Penn

Sometime it pays to be different. Every professional copywriter and art director should well know all of the caveats used in designing a print ad. *Don't* use a fine serif type in reverse; *don't* use fine-line artwork in an ad that also carries heavy blacks; *be careful* not to use heavy solid colors when delicate halftones are used in four-colors, etc., etc. New York and Pennsylvania, a Pulp and Paper manufacturer and a subsidiary of Curtis Publishing, wanted to get production and graphic arts people to pay attention to the outstanding printing capability of its paper. Aldo Santi and Dick Morrison of the old O. S. Tyson Agency and NY Penn's VP Marshall Dana decided the best way to proceed was to break all the rules.

This campaign, known as the torture test, was developed as a challenge to designers of world repute to create a design that violated their particular list of "no-nos" and in effect "tortured" the printing capability of the paper. Their designs were printed and run as inserts on the manufacturer's paper to demonstrate its printability.

Answering the challenge were designers of international repute, such as Herb Lubalin (shown), Ladislov Sutnar, Henry Wolf, Ray Kuhlman, Hans Hillman, Leo Lionni (of *Fortune* magazine fame), S. Neil Fujita, George Giusti, Paul Rand, Thomas Eckersley, Jacques Nathan-Garamond and others. Each produced their creation, and, in addition to the copy describing the torture test challenge, a picture and biography of each were included.

Run as inserts in graphic arts magazines of the 1950s, the ads won the highest readership ratings ever recorded in the then-popular *Printers Ink*.

Norton

Piet Mondrian was a Dutch painter and founder of neoplasticism, whatever that is. Before he died, though, he propounded the theory that "the logical outgrowth of painting is the use of pure color and straight lines in rectangular opposition." In other words, he was big into horizontal and vertical lines. A number of art directors in the late 1940s and early 1950s used Mondrian-like layouts to perk up readership and, in many cases, it was unusual enough to lure straying eyes to a page of copy. A recent case in point is this Norton ad which combines pure squareline work with some excellent color photography in a way that would make old Piet jump right out of his neoplastic socks. "Safety In Numbers" is three-quarters illustration and one-quarter copy, but you don't even have to read the copy to get Norton's message. Norton is the leading supplier of industrial safety products; you'd better believe it.

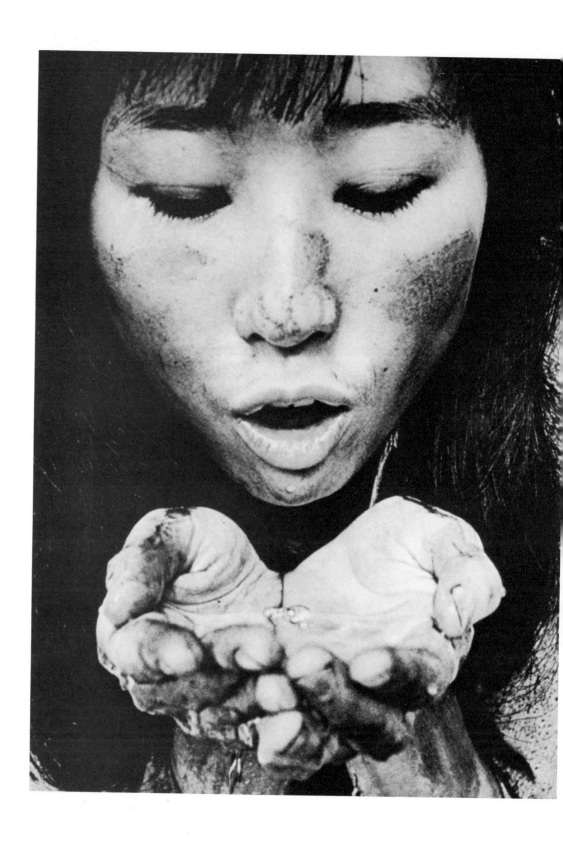

Olin

Here's one of the greatest corporate ads of all times. It ran in 1964 just prior to all the acrimonious anti-Viet Nam feeling that pervaded the United States. Key to its success was a stock photograph selected by Bill Taubin, who was then an art director at Doyle Dane Bernbach. But the scintillating copy by Chuck Kollewe doesn't pall in its reflection by any means. The body text very quickly points out that tons of Olin chlorine is being shipped to the Vietnamese to help stomp out typhoid and other diseases related to water-contamination.

In Viet Nam jungle water is more dangerous than the Viet Cong.

If that fact hasn't made headlines, it's because it isn't news. Contaminated water has been killing Vietnamese for over 2,000 years.

Typhus, typhoid, amoebic dysentery, infectious hepatitis—these are just a few of the deadly assassins.

Coupled with malnutrition, they limit the life span of a peasant to 35 years. And they guarantee that more than half his children will die before they're 5 years old.

With statistics like these all too common in underdeveloped areas, it's obvious that the ultimate war to be fought in Viet Nam and southeast Asia will be the war against disease.

Fortunately, it's already underway. Olin, for example, is shipping tons of chlorine to Viet Nam each month for on-the-spot water purification.

Not ordinary chlorine, but a special dry granular type called HTH,* developed in this country by Olin back in 1928. It's easier to handle and safer to use. So safe, our G.I.'s use HTH tablets to purify water right in their canteens.

And when more sophisticated means are used to purify water, other Olin products will find use abroad.

Chemicals like biocides, to prevent the growth of algae from clogging lakes, rivers and reservoirs. Hexametaphosphate, to prevent rust scale from forming in pipes. And tripolyphosphates, to soften water.

Olin is even working on the development of large, portable chlorine systems for use in remote areas by small villages.

It's only a beginning, though. As the world population grows, pure water, like food, will become ever scarcer.

That's why in the jungles of Viet Nam, as throughout the world, the war against disease will go on long after man has made peace with man.

Olin

Olin is Chemicals, Metals, Squibb Pharmaceuticals, Paper & Packaging, Winchester—Western Arms & Ammunition.

The last paragraph is worth quoting because it says in very few but beautifully written words what kind of image Olin wants to convey: "That's why in the jungles of Viet Nam, as throughout the world, the war against disease will go on long after man has made peace with man."

According to Stefan Blaschke, then the account supervisor, this ad won every conceivable award presented that year, including the prestigious *Saturday Review* award as well as a special Rutgers University achievement award.

The strength of Fiberglas has helped boost the pole-vault record 2 ft. 10¾ in. How much stronger would it make your product?

In 1942, a pole-vaulter soared 15 feet 7¾ inches with a bamboo pole. And nobody could top his record for *fifteen years.* *

Then, along came poles made with Fiberglas.* Lighter. And *springier.* They literally catapulted vaulters to amazing new heights. Today's outdoor record: 18 feet 6½ inches!

Fiberglas reinforcement, added to plastic, can be engineered into a materials "system" with almost any desired characteristic.

Products like large-diameter pipe and motor-home bodies benefit from the same strength that improved vaulting poles. Other products like playground equipment and transformer covers are better because they're more moldable, electrically nonconductive, lighter, more durable, or less expensive.

Would a Fiberglas materials system make your product better? We'll help you find out. Write: D. L. Meeks, Owens-Corning Fiberglas Corp., Fiberglas Tower, Toledo, Ohio 43659.

Owens-Corning is Fiberglas

Owens-Corning

The four-color ad from Owens-Corning does everything right. We all know the Type "A" layout works well for readership. Also, the four-color action photograph is a real stopper. Research tells us that photographs using babies will attract more people than any other kind of illustration. (This is also true of horses, dogs and other animals and pets.) It's also true that sports-oriented photographs greatly appeal to male audiences. And this is exactly the target Owens-Corning is attempting to reach with their *Business Week* ad.

The creators of the ad were smart enough to put the name of the product in the headline. This is a good way to ensure that casual "glancers" or "thumbers" will know what brand, product or company is being advertised. Again, the uppercase and lowercase headline, the 10-point minimum Roman body text and three narrow column widths all combine to enhance ad readership.

Peterbilt

What do you do when the market you once dominated starts to dwindle? Peterbilt had long been recognized as the Cadillac of the heavy-duty truck industry. Historically, Peterbilt trucks had been the pride of the owner/operator market. But when adverse legislation, coupled with a weak economy, shifted the market to fleet sales, Peterbilt's San Francisco-based agency, Pinne, Garvin & Hock, went on the attack. Creative director Robert Pinne and art director Pierre Jacot came up with this startling campaign of ads for *Business Week*. And they are simply perfect. Jim Blakeley, a

efficiency—an important plus in these energy-conscious times. What's more, our customers attest to the fact that our cab construction, coupled with matched components, can provide a cost-effective means of transportation. And whether you own one truck or a whole fleet of trucks, we've got a Peterbilt designed to meet your short and long-term needs.

Even more important from a long-term investment standpoint, more than 92% of the 1968 model year Peterbilts originally registered were still working at last count.* That long productive life cycle is why we call Peterbilt the Money Machine—while it saves you money, it can make you money. Like heroic achievement, Class is never outdated. For our free booklet,

The Investment in Class, write: General Sales Manager, Peterbilt Motors Company, P.O. Box 404, Newark, CA 94560. *Source: R. L. Polk & Co.

A DIVISION OF PACCAR

super photographer, put together a cast of thousands (well at least 34 models, count them) for this stunning shot of an action-packed ticker tape parade. Just think of the logistics involved in this one—not to mention the street cleaner's bill. There's no way you can avoid stopping in mid-*Business Week* to read this fine message. According to Ed Hughes of *Business Week*, this series is one of the all-time highest scoring campaigns ever to appear in his magazine. Copywriter Jeff Patterson's five columns of prose ends with a quality encomium that bears repeating: "Like heroic achievement, class is never outdated." Nor is good hard-selling advertising like this.

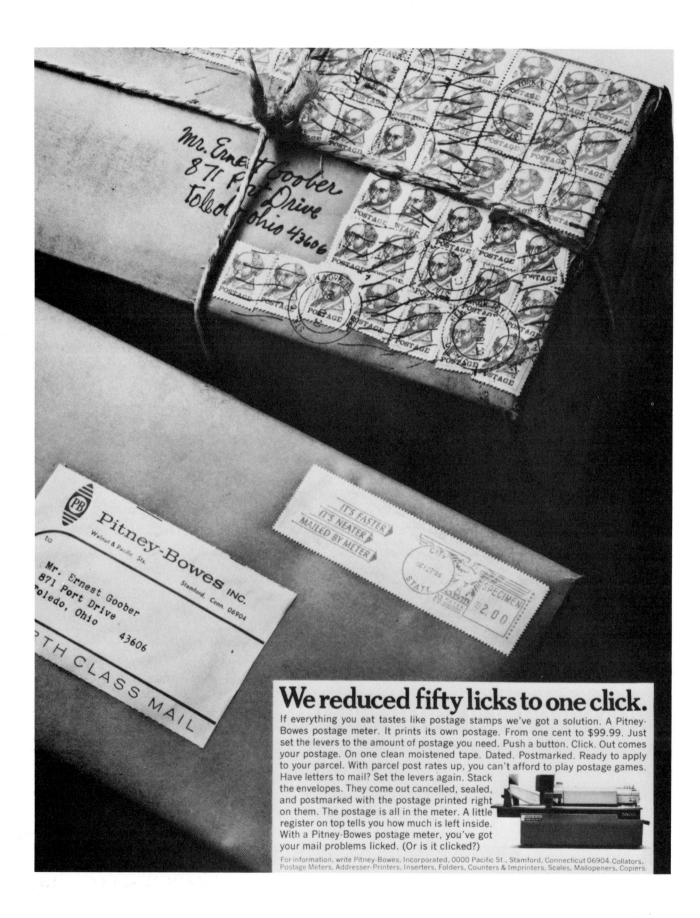

Pitney-Bowes

Industrial Marketing chose Richard K. Jewett as Adman of the Year in 1967 primarily because of the excellent print campaigns he spearheaded for Pitney-Bowes over a great number of years. IM's Copy Chasers made their selection based on "Kirk" Jewett's fulfillment of his company's objective of strengthening its image of "proprietory ownership" of products and services in the postage meter and office equipment field. Gallup & Robinson office equipment ad studies show that Pitney-Bowes' ads score almost twice as high as the industry norm. For over 20 years, P-B ads have not only scored number one in the field, but their grades have been so high they have raised the overall category average. This means that each new Pitney-Bowes advertisement that is studied has to compete against a higher median overall. Caused by themselves! Ironic, but true.

The headline of this greatest ad, "We reduced fifty licks to one click," done in a neat, bold uppercase and lowercase typeface, whets the reader's appetite for the message to come. So does the before-and-after photograph of the packages addressed to Mr. Goober. The copy starts out in terse, staccato-like sentences: "If everything you eat tastes like postage stamps we've got a solution." We defy you not to read the rest. Truth is you have to. This is engagement at its best. Also, notice the body copy is set in 10-point, easy-to-read size in order to cater to the audience's reading eye. And the send-off is terrific: "With a Pitney-Bowes Postage Meter, you've got your mail problems licked (or is it clicked?)."

Congratulations, Kirk.

"Polaroid" and "Polacolor"® *Average time in metropolitan areas.

Polaroid

You have to hand it to copywriter George Rike and his art director Lew Byck of Doyle Dane Bernbach. They really put together a stunning advertisement for Polaroid. It's hard to beat a comparative ad when the advertiser is comparing the reproduction ability of its product exactly as it was reproduced on the printed page. The image on the left was printed from a transparency that took eight hours to process. The one on the right is a reproduction from a Polaroid print that was developed in 60

THE 8x10 POLAROID PRINT— 60 SECONDS.

Now, with Polaroid's instant 8x10 film you can get original art for reproduction in only 60 seconds. Superb quality photographs with color, sharpness and detail that can be reproduced on the printed page using any reproduction method.

And, as you can see, the Polaroid print delivers results comparable to a transparency. Because Polacolor 2 prints have density ranges that closely match those possible with 4-color printing processes, everyone can judge *ahead of time* what the final result will be. And if necessary, prints can be retouched.

Art directors and photographers also have new flexibility and creative latitude *during* the shooting. And know instantly they have exactly the shot they need. There's no waiting. No costly retakes. No surprises.

So with Polaroid 8x10 prints, you get quality at a cost comparable to that of a transparency. But in just 60 seconds.

For more information about Polaroid 8x10 film, contact a Polaroid Professional Products Dealer, or call us toll free: 800-225-1618. (In Massachusetts, call collect: 617-547-5177.)

POLAROID
INSTANT 8x10 COLOR PICTURES

seconds. What a dramatic way to tell a story. The ad did the job it was essentially designed to accomplish. Up until it ran, sales for 8 x 10 instant print film were very sluggish. For almost a year after its introduction, photographers were still leery about the product's image quality. Then this ad ran in *Professional Photographer, Print* and *Photo Methods*. Leads immediately started pouring in. Ultimately, this great piece of copy brought in a total of 3,000 sales leads which were directly responsible for over 200 sales. Some great job!

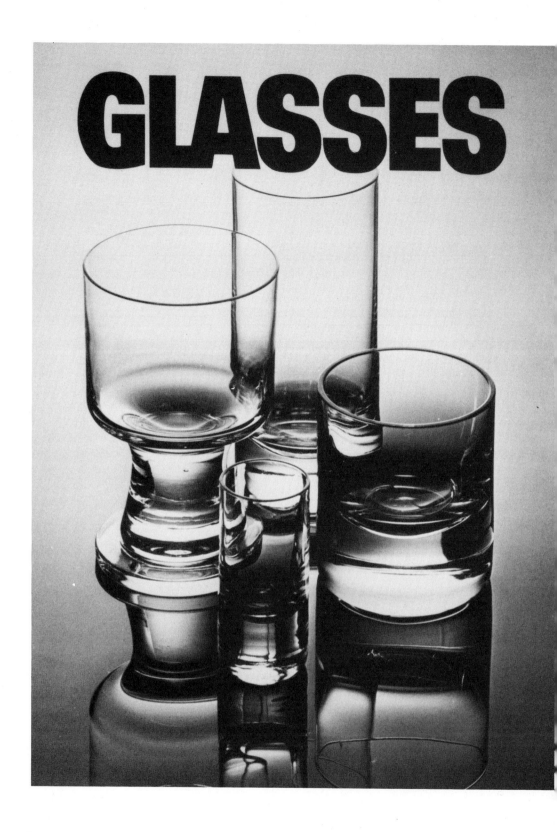

PPG Industries

The greatest line in the movie *The Graduate* was the bit of career advice given to Dustin Hoffman by his father's well-meaning friend: "I've got just one word to say to you, boy: *Plastics!*" And that's exactly what this PPG Industries advertisement is all about: plastic lenses, not glass lenses. Pittsburgh's Ketchum, MacLeod & Grove was the agency that produced this very strong appeal to opticians, ophthalmologists and optometrists. And they don't banter. The intro gets right to the heart of the matter by telling them and instructing them

that "the difference is already clear. Which is why many of your own patients who used to ask for glasses are now asking to see Plastics, fashion eyewear with optical plastic lenses made from CR-39 monomer from PPG."

There was no coupon to sign off this fine ad. Only a last-sentence request to write PPG Industries, Inc. And did they ever! According to the advertiser, more than 4,000 requests for additional information were generated by this campaign. Our hat is off to Scott Young, the copywriter, and to Joseph Kravec, KM&G's art director, for this beautifully crafted advertisement.

Comfort in Safety

THE automobile manufacturer builds into his particular car those units which he knows will give satisfactory service for the life of the unit. The brakes are designed to provide you with "comfort in safety", but only so long as the brake lining is in *perfect condition*.

The manufacturer does his utmost to assure safety. Yet, *his* responsibility is limited. After the car is in *your* hands, the responsibility rests upon you. Brakes need occasional adjustment and, in time, brake lining will become worn and require renewal.

To obtain *the same measure of safety* which the manufacturer intended should be yours, you must have the brakes inspected and adjusted periodically.

Your garageman can readily render this service, and if new silver edge Raybestos brake lining is needed he will apply it by the Raybestos Method which means— "COMFORT IN SAFETY!"

If you will forward the coupon, we shall be glad to tell you the name of the serviceman in your vicinity who will inspect, adjust and reline your brakes with Raybestos by the Raybestos Method.

THE RAYBESTOS COMPANY

FACTORIES: Bridgeport, Conn. Peterborough, Ont., Canada
Stratford, Conn. London, England

BRANCHES:
Detroit, 2631 Woodward Avenue . San Francisco, 835 Post St.
Chicago, 1603 South Michigan Avenue

THE RAYBESTOS COMPANY, BRIDGEPORT, CONN.

GENTLEMEN: Please send me your booklet "BRAKES—Their Care and Upkeep," and the name of the nearest serviceman who will inspect, adjust or reline my brakes by the Raybestos Method.

Name _____

Address _____

I drive a _____

Raybestos

Often an industrial company that supplies original equipment manufacturers with parts or services has to run corporate ads in consumer publications to plug its customers' products. Done right, this can bring lots of respect (and hopefully more business) to the advertiser. In the early 1920s, Raybestos ran this Norman Rockwell campaign in the *Saturday Evening Post* (where else?) to tout its brake shoes. In those days, Rockwell illustrations were on most of the *Post's* covers, so using him to do the art for a Raybestos ad was a natural tie-in. Notice that the short copy was chock-full of kudos for automobile manufacturers. Phrases like "the manufacturer does his utmost to assure safety" abound in these ads. Raybestos not only plugs its customer, but also asks its customers' customers to look into Raybestos brake linings when their original brakes wear out and have to be replaced. These sensational ads also include a coupon which gives Raybestos an opportunity to write inquirers and tell them where they can go to get service. This enables Raybestos to push its stocking service stations. So, in effect, the ads are just as beneficial for OEM's as they are for replacement sales.

Norman Rockwells' genius for capturing Americana (and therefore readership) is to this day still unsurpassed.

Raybestos currently sends out a series of four of these 50-year-old advertisements, suitable for framing, to select customers in appreciation of their loyalty.

Beech, with its colorful past and bright future, is now part of Raytheon

These two airplanes—one old, one new—characterize better than anything else a company that holds a unique position in American aviation...Beech Aircraft Corporation.

The classic Beechcraft Staggerwing biplane, initially produced in 1932, was the company's first aircraft. It remained in production for 14 years, winning a reputation for speed, dependability, and a touch of elegance—qualities that have characterized Beech aircraft ever since. More than 200 Staggerwings are still in existence—handsome and durable reminders of a proud past.

Today, Beech produces 23 aircraft models serving the entire general aviation market, ranging from light, single-engine trainers to the King Air series of business aircraft.

The Beechcraft Super King Air 200 (shown in the background above), with its fuel-saving turboprop engines, is truly an airplane for the times. It has, in fact, outsold all other competitive turbine-powered aircraft for the fourth straight year.

Current production of the Super King Air continues at a record rate.

Raytheon

For years, Raytheon corporate advertising has been outstanding. Beautiful four-color spreads with simplified formats do a fantastic job of spreading the word about Raytheon's plants, products and people. The main reason this campaign works is because ad director Van Stevens insists on superb four-color photography. There's no room for second-best in his lexicon. And that's really what good advertising is all about. A grabber of an illustration or main photograph is 90 percent of the battle. It's corny but true that a good photograph is worth a thousand words—especially here.

Raytheon's agency account supervisor at Creamer Inc., Cliff Gulbransen, is a veteran ad maker who has been behind some of the company's greatest ad efforts. He and his creative team, art director Sal Venti and copywriter

As the cost of fuel increases, and commercial airlines continue to reduce service to all but the major cities, efficient business and commuter aircraft will play an ever-increasing role in air travel. Beech is in a strong position to serve this growing market, not only with the King Air series but with the Commuter 99, soon to enter production, and the new Model 1900, 19-passenger, pressurized commuter currently in full-scale development.

Now Beech is part of Raytheon, adding an important new dimension to our size, diversity, and potential for growth.

Raytheon...a company with expected 1980 sales of $4.9 billion in electronics, aviation, appliances, energy, construction, and publishing. For copies of our latest financial reports, please write Raytheon Company, Public Relations, 141 Spring Street, Lexington, Massachusetts 02173.

John Williams, have a commitment to excellence that shows through in this Raytheon ad which features flights of an old and new Beech aircraft. Nostalgia always pays off at the readership bank, and, when you combine it with modernity, just watch those Noted scores hit the ceiling. Over the years, this campaign has consistently won top-of-book readership awards. And it's as flexible as can be. This particular ad announces that Beech Aircraft was absorbed by Raytheon. When it first appeared in 1980, it attained a gigantic 76 Noted score. Another in the series touts a product, still another sells a service, but they all end up an integral part of a strong campaign telling the story about Raytheon, the $4.9 billion company in electronics, aviation, appliances, energy, construction and publishing. It's an ideal way to spread the word that the corporate-conscious Raytheon wants to convey.

IMMEDIATELY AVAILABLE

TACTICAL WEAPON SYSTEM

Only Mach 2 fighter-bomber w/heavy mil exp, capable ground strikes ceil zero nite/day. Fully integrated navig, fire- and flight-control syst. Pres empl deterrent specialist USAF. 24 hrs 7 da/wk. Verstl, prov abil counterpunch w/ nucl or conv weapons, support ground forces, tk chg any mission, any weather, any time, anyplace. Avail NOW. F-105D Thunderchief, Republic Aviation Corp., Farmingdale, L. I., N. Y.

Republic Aviation

It's hard to appreciate this advertisement as shown in its present form. It was a shocker in 1963 when it appeared just once in *Time*. It was centered and floated on a full white page—but there was a slight yellow tint behind it to give an aura of a classified ad. Even in those days, the media buy was very expensive for a company mainly interested in advertising in aerospace and government trade journals. Let's face it, you can buy a lot of pages in *Aviation Week* for the cost of just one insertion in *Time*. This really took guts on the advertiser's as well as the agency's part.

The idea behind the ad was to subtly let the governmental opinion molders (congressmen, senators and the military) know that Republic was still around the corner, ready, willing and extremely capable of supplying additional Thunderchiefs should the need arise. The ad also contained a considerable amount of puffery about the quality of these fighter-bombers that were being paid for by the U.S. taxpayers who also happen to be readers of *Time*.

Apparently, *Time* gets considerable exposure overseas, because Republic was contacted by some governmental "biggies" from the Saudi Arabian Council which specifically asked for price and delivery information. The Republic ad was actually clipped to the Saudi Arabian letterhead. Unfortunately for Republic (and fortunately for a number of other Mideastern countries), the United States prohibited Republic from sending the Sheik a bid. Credit for the idea and copy for this particular "great one" goes to John deGarmo, formerly of deGarmo Inc. Ken Ellington was the ad honcho from Republic at the time.

Rockwell Report

by W. F. ROCKWELL, JR.
President
Rockwell Manufacturing Company

THE POWER of an unethical minority to smear the entire industry or profession of which it is a part has, unfortunately, been demonstrated once again in the exposé of the rigged TV shows.

It is the public, of course, which suffers most directly by this outright deception. But injured along with the public are the thousands of advertisers and advertising people whose standards of decency are second to none. In the minds of many, including some politicians, this majority is being classed with the minority who grab for the quick dollar either through outright deception or with ads that certainly violate both good taste and credibility. Even more seriously, the entire economy is injured, too, because disbelief in selling messages cripples the selling process on which our economy rests.

Selling any product consists of telling people what it will do for them. The early craftsman told his neighbors about his wares, and they bought. If he didn't lie to them about his product, they bought again and again, and his business grew. Advertising is merely an extension of that simple process, a modern day mechanized method of telling more and more people at lower cost per person.

But nowhere along this path of business evolution has the essential principle changed: If you deceive your customers, they won't buy a second time. If disbelief in your word spreads, most people won't even buy the first time. And nowhere is this truth more binding than with industrial buyers, such as those we serve, who are so knowledgeable in their fields and so sophisticated as buyers that high standards of credibility are mandatory.

Most businessmen know this. They go to great lengths to protect the integrity of their products and of their names. And they have enough healthy respect for the intelligence of their customers to know that it's business suicide to try to fool even a few of the people some of the time.

* * *

Here is, perhaps, the best kind of evidence that it pays to play fair with customers. It is a letter from a man who had just bought two Delta Tools for his home workshop: "Warranty cards enclosed. I'm not much concerned with the warranty on these new Delta Tools. They replaced a Delta combination saw and jointer which I bought in about 1932 and it is in just as good condition today as when I bought it. I gave it to my son-in-law when I bought the new 10 inch saw and 6 inch jointer. I've had the same good service from my Delta Band Saw, Lathe and Drill Press which were bought in 1932 and my sons-in-law will probably inherit those in good condition. I replaced my 1932 Delta "jig saw" with a new model about 10 years ago and added a Delta shaper and grinder at about the same time. They have been equally satisfactory.

"It seems to follow that I am unconcerned about your one year warranty because my Delta Tools have so far lasted as long as my marriage, 28 years, and my wife is still piling stuff on my work bench to be fixed."

* * *

A new high pressure water meter, the Model 504 Rockwell "Five Pointer" designed for oil field water flooding, has been introduced by our Petroleum and Industrial Division. Water flooding is a widely used technique for forcing oil out of the ground by pumping water into the ground. It requires very accurate metering so that the amount of water pumped underground, and therefore the force exerted by it, can be controlled and kept in balance.

This is one of a series of informal reports on the operations and growth of the

ROCKWELL MANUFACTURING COMPANY
PITTSBURGH 8, PA.

for its customers, suppliers, employees, stockholders and other friends

R-6001

Rockwell

Everybody in the business should remember this great corporate campaign. It ran for years under the banner "Rockwell Report" and was signed by Rockwell's president at the time, W.F. Rockwell, Jr. The series of 2/3-page, black and white, long copy advertisements appeared in publications like *U.S. News & World Report, Business Week, Newsweek* and the *Wall Street Journal.* And they must have done the job they were intended to do. Today, Rockwell International has annual sales of well over the $6 billion mark and employs over 100,000 people.

The key here is Mr. Rockwell himself. Here's a guy who believed so much in advertising that he was willing to put his name on it. *Industrial Marketing's* Copy Chasers thought so highly of Mr. Rockwell and his ad campaign that they named him Industrial Adman of the Year in 1959. And well they should. It isn't very often you can find a champion of advertising among top industrial management people.

The Rockwell Report does a great job of showing the public what the company's chief executive officer's personality is all about. The idea here is that it's easier to identify what a man stands for than what an inanimate company stands for. The ad copy gives some very stimulating management viewpoints and features newsy things like new products, success stories, available literature and the like.

The campaign is designed to build confidence in the man, which leads to confidence in the company, and then quite logically builds confidence in their products.

In order to keep the series from being dull, the Marsteller ad agency art directors kept changing Rockwell's photo (the only halftone in each ad). They also used different body text typography in order to help alleviate the boredom factor. But the different photos were the only element in the format which changed.

This superb campaign was a precursor to many of the op ed ads we see today.

We'd like to spring something on you. Free.

84 pages of practical spring design information.

Includes materials selection data. Plus formulae for some specific applications.

Kinds of springs covered: Compression, extension, torsion, flat torsion, motor, retaining, stampings and clips, Belleville and wave washer.

...engineering terms and symbols to help you if you want to design your own springs.

And if you can't get all the help you need from our engineering book, write for one of our engineers. We have the people and the know-how to devise, design, develop, and deliver to meet your needs. Exactly.

For your copy of the book—or design help, write to Automotive Products Division, Clifford at Bagley, Detroit, Michigan 48231. And watch us spring into action.

North American Rockwell

Rockwell Spring

Here's an absolutely fabulous success story. This Rockwell ad entitled "We'd Like to Spring Something on You. Free" proves again the pulling power of key words. The "Free" bit can really work wonders. What it did in this four-color ad that ran in *Design News* in 1972 was to get the best advertising response in the publication's 26-year history. The magazine had run more than 74,000 pages of ads during those 26 years, and the Rockwell ad, which offered a free 84-page mechanical spring design guide, pulled 3,038 inquiries from a single insertion! This doesn't even count the direct responses Rockwell received by letter and telephone that were also a direct result of the one ad's one appearance. Ralph Watts, Rockwell's former manager of communications, and his agency, Campbell-Ewald, share the "greatest" honors on this one. Elmer Mellebrand was the C-E art director, Joseph Paonessa was the writer.

Rome Wire

Here's an oldie that was part of one of the very first institutional advertising campaigns ever run by an industrial company. "Out on the copper highways" ran in 1926, and it has the honor of having won the first industrial advertising award granted by the Business School of Harvard University. Moser & Cotins of Utica, Rome Wire Company's ad agency, was responsible for coming up with the insert idea which was designed to position Rome Wire as the undisputed leader in the copper wire manufacturing field. A market research project revealed that Rome Wire was the country's most extensively used brand, but because of the rapid growth of the wire and electrical industry, many people were not at all familiar with the company and its enormous facilities.

This very dramatic insert, using striking four-color artwork of a linesman climbing a utility pole, was a real breakthrough when it first appeared in magazines such as *Electrical World, Coal Age, Electrical Railway Journal, Electrical Merchandising* and the like. For one thing, it used a consumer copy approach in order to romanticize a rather prosaic product to the readers of these publications. The copy starts out like a novel: "In blazing sun and blinding blizzards, the bronzed guardians of service carry on. Theirs is the responsibility of construction and maintaining service on the copper highways of the country." It goes on to describe how for 20 years the company had grown from a small wire shop to mills that covered 20 acres of manufacturing floor space. The entire effort featured Rome Wire as the "Bronzed Guardians of Service," and it did it well. Both Rome Wire and its agency received an honorarium of $2,000 each from Harvard for its first-place industrial award. You have to believe that $2,000 was worth a lot more back in 1926. How about $200,000.00 today?

The Steel Strike and Ryerson Steel Service

As these lines are written * the strike announced by the president of the CIO United Steelworkers on October 1 is still in effect.

Naturally, the duration, extent and consequences of this stoppage in steel production are unpredictable. But for the sake of our national economy and the welfare of all we hope for a speedy and satisfactory solution.

Fortunately, during the past few months, the condition of Ryerson steel inventories in our

* Friday, October 7.

plants has improved materially. Orders for high quality carbon, stainless and alloy steels are being filled promptly. And during this emergency they will continue to be filled, at our established prices, while our stocks last.

Likewise continued will be our long established policy of providing prompt, personal service whether an order calls for a single bar or a carload. So contact your nearest Ryerson plant for any steel requirements—and we will do our level best to take care of you.

Joseph T. Ryerson & Son, Inc. *Plants at:* New York • Boston • Philadelphia • Detroit • Cincinnati • Cleveland • Pittsburgh • Buffalo • Chicago • Milwaukee • St. Louis • Los Angeles • San Francisco

RYERSON STEEL

Ryerson

With firm hand and steel-trap memory, 87-year-old Keith Evans (one of the Business/Professional Advertising Association's founders) filled me in on this great advertisement for Ryerson Steel. "The Steel Strike and Ryerson Steel" was written by Keith over 30 years ago. That was about the time that Harry Truman was threatening to take over the country's steel mills and all industry was going ape. Ryerson was smart enough to go the all-type route and very dramatically editorialize its message in an almost funereal style. Taking advantage of the news opportunity, the copy beautifully telegraphs the punch of the message. I defy anyone not to read further than this first sentence, "As these lines are written* the strike announced by the president of the CIO United Steelworkers on October 1 is still in effect." Note how the asterisk leads your eye to the line "Friday, October 7" at the bottom of the left-hand column. This is another great device that guarantees superb readership. The timing was also perfect. The ad appeared just a week later in the October 15 issue of *Newsweek* and guess what—it scored second in total overall readership and was only topped by a much more expensive, four-color ad. We sure don't see too many all-type ads like this anymore.

Nor do they make too many all-time great advertising people like Keith J. Evans.

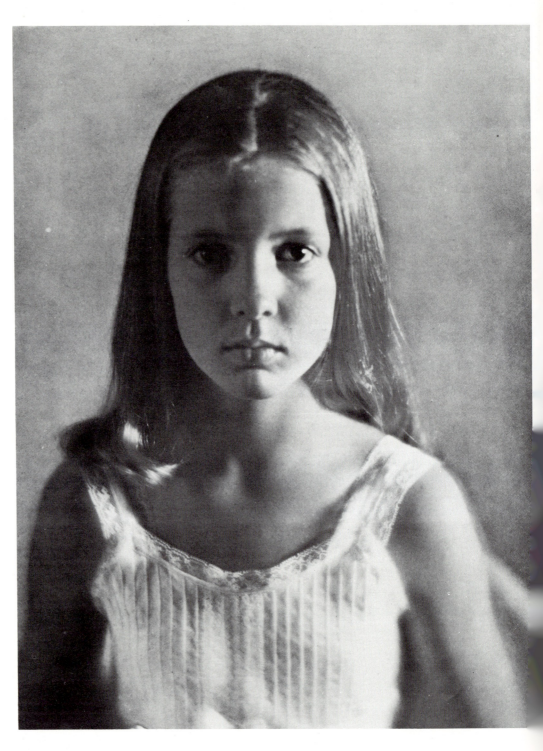

Sears

For years, Sears Roebuck had the image of a chain of schlock stores that sold a lot of cheap merchandise. Back in mid-1961, George H. Struthers, Sears VP in charge of merchandising, launched a corporate campaign in major American magazines that was, as he put it, "designed to give present and prospective customers a new view of Sears—our people, our policies, our merchandise, our service." It also began to switch emphasis away from the name Sears, Roebuck & Company to just plain Sears.

The greatest of all in the series was the one Ogilvy & Mather's then copy supervisor Bob Pasch came up with. He saw a stunning photograph of this young child in a photographer's sample case and decided he just had to build a Sears ad around it. The photographer tried to duplicate the shot with another model in a

How Sears helps your daughter choose her first bra

Bring your daughter to Sears, Roebuck and Co. for her first bra and girdle. Sears figure experts are trained to select the correct garment for every woman. But they take *special* pride in starting a youngster out right.

REMEMBER the day you bought your first bra? Was it an exciting, grown-up kind of day? Or awkward and embarrassing?

Sears takes great care to make sure its young customers remember this day with pleasure. Sears bra and girdle people know how to make your daughter feel at ease. Many of them are *graduate fitters*. This means they have completed Sears highly comprehensive figure-fitting training course—and passed a stiff written and oral examination.

Nobody asks, "What size?"

Your daughter's first step for her first bra at Sears, is into the fitting room. The Sears graduate fitter will keep her measurements on file at the store—and bring them up to date as she grows, and as her measurements change. This written record lists her size, figure type, style of bra and the fitter's comments.

But this is by no means a service for young girls only. When *you* go to Sears to buy a bra or girdle, the Sears fitter can measure you as carefully, too. And when you choose a Charmode bra or girdle at Sears, you can be *sure* of its quality and workmanship. For example, the elastic in the Charmode Cordtex® bra wears *longer* than any other elastic tested in the Sears laboratory.

Free alteration—on the spot

If a garment doesn't fit you *exactly*, it must be altered. Often the Sears fitter goes right to her sewing machine and makes darts and tucks on the spot. Free.

Other alterations are free, too. Taking in the hips of an "all-in-one" for a woman with a full bust and small hips. Placing flannel strips inside a girdle for extra comfort. Changing the position of garters. Special fitting of surgical and maternity garments.

Sears fitters know every woman can have a better figure—with a bra or girdle that *fits*.

The Sears way of doing things

This kind of professionalism is the Sears way of doing things. You find it in *all* departments—and in all Sears *people*, from graduate fitters to home decorators to TV repairmen. And *every* salesperson in *every* one of the 740 Sears stores.

It is their finicky attention to detail that makes Sears a "customer's store"—and lets Sears offer its famous promise: *Satisfaction guaranteed or your money back.*

← You'll find Sears Charmode bras and girdles in Sears stores and in the Sears catalog. You can't get better quality for the money. The same thing goes for girls' slips and petticoats. The slip this girl is wearing is just $1.98. Remember, you can always charge it at Sears.

Sears slip but couldn't get a picture that captured the quality of the original. George Struthers then saved the day by ordering his vendors to *copy* the slip. They did and the Ogilvy team built an ad around this fantastic photograph with its simple look of innocence. Pasch wrote the headline and Leah Thuna, a star Ogilvy copywriter and budding playwright, wrote the copy. The factual copy was designed to convey an aura of authority and importance along with editorial credibility. That's why it was written like an editorial.

The ad ran in 1963 and the results were instantaneous. A deluge of complimentary letters flooded in, accompanied by a few from concerned parents who were worried about rushing their daughters into maturity before their time. But the overall objectives were achieved—and then some! It's certainly one of the few times that a corporate advertising message has sold enough products (slips and bras) to pay for itself.

Shell Answer Book #1

THE EARLY WARNING BOOK

How you can spot some car problems before they cost big money

By Russ Russo, Shell Car Care Expert

Come to Shell for answers

Shell Oil

The energy crisis started in the early 1970s and was exacerbated in 1973 by the oil embargo. And who took the rap? The full wrath of the citizenry fell on every oil company operating in the United States. Shell research at that time indicated the public (every adult 18+ years of age) would not believe anything claimed or promised by an oil company. In order to blunt this sharp knife of criticism, Shell met the problem head-on with a corporate advertising campaign called "Come to Shell for Answers." Launched in the spring of 1976, the campaign's objectives were to enhance Shell's reputation for responsible behavior, contribute to its credibility, and support brand-name reputation for its products, services and people. Shell also realized it could stay one step ahead of competition if it could prove its responsiveness to its customers' needs by providing beneficial tips on the care and handling of the family car.

Since the inception of the campaign, the company has prepared 27 fact-filled eight-page booklets which have been distributed through national magazines and Shell service stations.

The books all follow the same format. They're each sequentially numbered and run off in handsome four-colors. The typography is excellent. Boldface questions are set in easy-to-read, large uppercase and lowercase type. The books also contain easy-to-fathom diagrams and clear, concise captioned photographs.

The proof of the pudding is unbelievable. Total company awareness since the campaign's beginning rose from 20 percent in 1976 to 66 percent in 1981. Ninety-four percent of the people who have received the booklets find them very or somewhat helpful. Research shows that Shell advertising is considered more favorably than all the competitors! And (get this) more than 900 million booklets have been distributed, and 960,000 letters requesting booklets and offering favorable comments have been received.

Their series has won an Effie, an Addy, a *Saturday Review* Distinguished Advertising Award and a *Marketing Communications* Marketer of the Decade accolade.

SKF puts its best foot forward

You're taking a step in the right direction when you call in the SKF man. First—you are enlisting the aid of a skilled engineer who is eager to apply his broad background and experience to your particular anti-friction problems. Second—he's the man that has a complete line of bearings available (more than 3000 sizes in the four basic types). Need convincing? Just call the nearest SKF office—and watch him prove it!

7823

Spherical, Cylindrical, Ball, and *Tyson* Tapered Roller Bearings

EVERY TYPE—EVERY USE

SKF®

SKF INDUSTRIES, INC., PHILADELPHIA 32, PA.

SKF Industries

Suppose you sell a product in a market loaded with competitors, each of whom manufactures to industry standards. Also suppose that prices are within mills of each other. And, to top it all off, you are a foreign-owned company and forced to bid against the "Buy American" legacy that's preached to purchasing agents since birth.

This was SKF's sales situation back in 1958 when art director Ken Sekiguchi and Herb Harris, copy chief of the old G.M. Basford agency (now Creamer Inc.), came up with this great eyepatch campaign (remember the Hathaway shirt ads?). In this case the eyepatch was a beautiful St. Bernard dog. The St. Bernard was used as a symbol of reliability and dependability, carrying with it a suggestion of coming to the rescue. And—dogs also have great attention value.

First indicators of success came from the field when salesmen reported being greeted with barks and "bow-wows" from purchasing agents. Letters also came from customers and prospects asking how the dog managed to carry several hundred pounds of bearings around its neck. (One engineer-reader actually computed the weight had the boxes been filled.) This fine campaign ran for six years and through four dogs. Customers actually nicknamed the dog "Skiffy", although the name was never used in any of the advertisements. Over the years, the ads consistently pulled the best Noted ratings in every measured readership study. Ancillary use of the campaign (and the dog) was highly beneficial. Reprints were used successfully in direct mail pieces, posters, and at trade shows.

A tremendous amount of publicity was generated as well. McGraw-Hill Publishing Company held a retirement party for the last Skiffy which was attended by famous dogs from Greyhound, Mack, RCA, Gaines and a dozen others, and I'm told officials from the ASPCA outnumbered the guests.

How St. Regis made a two-by-four that's

The world's tallest tree, shown on the right, is a 368 ft. redwood in a national forest near Eureka, California.

And yet we made a two-by-four that's 400 feet long—32 feet longer than the world's tallest tree. Forty stories of continuous two-by-four.

How did we do it? Simple. We glued 60 off-sized pieces of two-by-four together with hot-glued finger joints like the one shown at the far right. And then got fifty men to carry it out into the field below and intentionally break it against a tree. Our tests show that the joints are stronger than the surrounding wood. And, once again, as you can see, it broke on the wood and not on the joint.

Theoretically we can make one any length.

Tops in two-by-fours.

Of course, our specialty is not 400 footers but eight footers—the so called stud, the main support of the home building industry. We're one of the largest makers of eight foot two-by-fours.

And, as in any operation where the raw material is non-uniform, there's a certain amount of waste. That's why we go out of our way to use the short leftover pieces. It's our policy to use the resource wisely. So we're out to use all the log—100%.

Everything including the smell.

One way we have of using the log more efficiently is computerized cutting. A scanning device sizes up the log as it enters the sawmill and our computer determines how to cut it to get the most out of it.

We figure these new computerized sawmills

St. Regis Paper

St. Regis Paper Company copyrights its two-page, four-color corporate advertisement and we don't blame them at all. Pardon the pun, but the photographer went to great lengths to show "how St. Regis made a 2 x 4 that's 32' longer than the world's tallest tree." It would have been so easy and far less expensive for the company to insist on an art illustration rather than actually fabricating a 400' long 2 x 4 and then photographing it being carried by a cast of 50 men and women on location.

But it would have produced only an arresting ad—not an engaging one. The photograph

32 ft. longer than the world's tallest tree.

will make the studmill as efficient as any modern manufacturing process.

Of course in our pulp mills we chemically break down the wood, so there we use virtually everything in the tree. Including the chemical components of the wood that give the tree its odor. They go into all kinds of chemical products.

Growing in all directions.

What's the future of wood building products? Are they being replaced by synthetics?

Pound for pound wood has greater compression strength than steel. And it's certainly more workable. Also wood is one of the more efficient insulators. A not inconsiderable fact in these days of energy shortage. And we're putting our technology into making the best possible use of wood.

Also we have a lot more building products than two-by-fours and dimension lumber. We make prestressed concrete walls and floors. Insulation and vapor barrier papers. And even culverts.

Technology and marketing.

In fact, this is the marketing stance of St. Regis toward all of our packaging, paper and lumber and construction products: to put the full weight of our technology at the disposal of our marketing effort.

Because, we at St. Regis believe the way to get ahead is to have the products of the future today.

The future belongs to those who stay up to the minute.

carries the reader to the actual scene of the advertisement. By involving the reader in the wood breaking experiment, St. Regis produced an ad that is not only an eye stopper, but also memorable. And by filming it for a television commercial at the same time, the company has come up with an adjunctive campaign in another medium. This is an excellent example of getting the most out of a single, great idea.

The corporate slogan, "The future belongs to those who stay up to the minute," obviously had the St. Regis copywriter and art director (both from ad agency Cunningham & Walsh) in mind when it was created. Who says industrial advertising has to be dull? Our hats off to the conceiver of this campaign.

Sonoco

This advertisement, run in the late 1960s, dominated *Fortune* Magazine every time it ran. At the time, Sonoco Products Company was the country's leading anonymous innovator in paper, plastics and metals. The anonymity was compounded by the confusion between the name Sonoco and Sunoco, the world's largest producer of petroleum. Rather than dream up a corporate campaign which would postulate its corporate conscience using the standard 3 p's approach (people, plants and products), the company opted to shoot the works by simply espousing its philosophy. This approach was so unusual that it got fabulous response and readership. The ad was a four-color, bleed spread with a dramatic photograph that dominated the left-hand page. The headline, "Sooner or later the company that ignores quality will end up where its products do," is

Sooner or later, the company that ignores quality will end up where its products do.

Nobody turns out junk because he likes to.
He does it because of pressure on prices.
But the way to deal with pressure is not to retreat. The way to deal with pressure is to innovate.
To find simple answers for complex problems.
To seek out new materials and new uses for existing materials.
To develop more efficient ways of doing things.
To really work at giving customers no less than they need, deserve and expect.
Now that you know how we do things, find out what we do. Write to us in Hartsville, S. C. 29550, for our free booklet.

Sonoco Products Company.
Innovators in paper
and plastics.

---

only great because that excellent junk yard photograph makes it great.

This is the case where one hand feeds the other. Photograph sells headline, headline complements photograph, and both combine to make a great ad. Short, succinct copy consists of only 10 sentences contained in eight paragraphs set in one wide column, Typographers would complain that the column was set too wide for readability, but because the type was set in a large 14-point, sans serif face, it is quite legible and rather easy to read. The copy platform is nothing more than a teaser. It makes a statement that would be absolutely un-American to disagree with, then asks the reader to write in for a free booklet which spells out what Sonoco does. R. Bruce Copeland, director of advertising and public relations of Sonoco, claims response was really good. They mailed out a heck of a lot of booklets, got some new customers, and made some new friends. That's what it's all about.

Spectra-Strip

Hal Pawluk at the agency bearing his name is the copywriter who wrote this Spectra-Strip spread and he's a sensational ad maker. This 1979 ad jumps right off the page and onto the reader's lap. Look at the mess of wires. Then read the headline, "We'll take it from here," and you've got to read the copy. And it's a nice piece of copy. It starts: "Let's face it, interconnections can really leave you at loose ends." This first sentence is also the copy's first paragraph. The second sentence is the copyblock's second paragraph and so on. The entire message goes for five sentences in five paragraphs. It's short, succinct and to the point.

That's really all that's needed because the Spectra-Strip logo really shows what the product is and does. Interestingly enough, the product tidies up that potpourri of jumbled wires in the main illustration. Before the campaign, Spectra-Strip hadn't advertised for five years. Recognition studies showed them at 1.6%; sales put them at number four in the market. The campaign ran one year and recognition jumped to a whopping 15.6%; sales were raised by 75% and pushed the company into the number two position in the market.

To top it all off, the campaign won several awards from *Sales and Marketing Magazine* and an elusive, hard-to-win ABP Objectives and Results award.

It starts out innocently enough.

A man tunes in a football game and tunes out his wife's attempts to be heard.

A woman gets so wrapped up in her problems she barely listens as her husband talks about his own.

And before long, without even realizing how it came about, a deadly silence starts to grow between them.

The fact is, listening, like marriage, is a partnership; a shared responsibility between the person speaking and the person listening. And if the listener doesn't show genuine interest and sensitivity to what's being said, the speaker will stop talking. And the communication will fail.

Which is why we at Sperry feel it's so critical that we all become better listeners. In our homes. And in our businesses.

We've recently set up special listening programs that Sperry personnel worldwide can attend. And what we're discovering is that when people really know how to listen (and believe us, there's a lot to know) they can actually encourage

Sperry

Back in the early 1970s Sperry had a fuzzy image. People were confused. Some referred to Sperry as the gyroscope people; others remembered the Sperry Rand Corporation or Sperry Univac, the computer manufacturer. A number of self-conducted surveys among investors, customers and prospects pointed out that Sperry was perceived as monolithic but not quite identifiable. The company's public image in terms of awareness and stature was not as strong and dynamic as it should have been. According to Richard Mau, Sperry's vice president of corporate and government relations, the company conducted still another bit of research. This one consisted of 18 months of in-depth interviews, the end result of which told them Sperry was respected for thoughtful attention to the suggestions, ideas and directions of its customers. In other words, Sperry was respected for listening. Given this gem of wisdom, Scali, McCabe & Sloves, Sperry's agency, came up with an overall theme for this

excellent corporate campaign. The line "We Understand How Important It Is To Listen" became the rallying point. The company pushed listening with all its might. Every ad promotes listening and ties Sperry to the idea. The most provocative of all these spread ads, "What's the Point of Talking, You Don't Listen To Me Anyway," shows an unhappy married couple silently staring out a window at opposite sides of their living room. The copy reads, "It starts out innocently enough. A man tunes in a football game and tunes out his wife's attempts to be heard." With an audience of some 70 to 100 million watching football on TV every Sunday you don't have to spell out the "Socko" effect of that prose. It really hits home, and that's probably why in the first year of the campaign over 125,000 Sperry listening booklets were mailed out in response to the small type offering at the bottom of each ad.

(Now—if I had only listened 10 years ago when my copy chief came up with the idea for a listening campaign...).

There are times when the heart longs more than ever to be home—and with no way to go. Koomey service men know, because they've been there, too.

FREE: An 11" x 14" full color lithograph of this original painting, suitable for framing, with no advertising message. While they last, and within Continental U.S.A. Write Sales Department, S&S Oilfield Division, P.O. Box 1473, Houston, Texas 77001.

Land & Subsea BOP Control Systems, Production Control Systems, Compressed Air Systems, Air Winches, Crown-O-Matic Safety Systems, Fresh Water Generators, Degassers, Tool Joint Locators, other systems and equipment.

STEWART & STEVENSON
OILFIELD DIVISION

Stewart & Stevenson

Every advertiser has at one time or another wanted to produce a Christmas advertisement for December issues. A lot have been done over the years, but there are only a few that have been memorable. This four-color single page Stewart & Stevenson Christmas ad is an outstanding example of what can be done when you use a fine illustration rather than a photograph. The artist has rendered a feeling here that no photographer could capture. The sheer loneliness of the two tool pushers decorating a tree made of metal tubes in desolate oil country engenders just the right feeling of pathos the company wishes to convey to the reader. The beautifully written short copy very quickly turns you on because at one time or another most everyone has been away from home during an important holiday. Let's face it, the ad gets you personally involved.

Jim Reid, S&S oilfield division's agency, merchandised the ad by making available free 11" x 14" full-color lithographs of the original painting without the advertising message for those who wrote in. A slew of requests came in on speed-memo forms, engraved letterheads, thank-you notes and Christmas cards as well. The ad only ran once (six years ago), and people continue to ask for free lithographs thinking they have seen the ad just recently. This is a fine example of how an emotional, heart-tugging appeal can do as well in industrial as it can in consumer advertising.

a vibro-energy* mill can grind materials finer than particles in this smoke

quickly and uniformly!

Smoke from your cigarette has particles as coarse as one micron—yet a SWECO Vibro-Energy Grinding Mill can reduce particles *uniformly* to sub-micron size! ✄ This is accomplished by efficient conversion of energy to particle size reduction through high-frequency, three-dimensional vibration. ✄ Materials such as titanium dioxide, zirconium silicate, calcium carbonate and kaolin are being milled in the Vibro-Energy Mill to give particles with maximum light scattering power, with a minimum of less effective oversize and undersize particles. ✄ Production of pigments with high opacity at low cost is accomplished by the Mill's capability for sub-micron grinding and close particle size control... often eliminating the need for costly classification. ✄ A full range of mill sizes is now available, 18" to 80" diameter. ✄ For catalog or free laboratory test on your materials, write:

SOUTHWESTERN ENGINEERING COMPANY
4800 Santa Fe Avenue,
SWECO Los Angeles 58, Dept. 129-435.

insist on the original vibro-energy grinding mill, finishing mill, separator — made only by SWECO

*TM SEC

Sweco

David Ogilvy would have exiled the creator of this Sweco ad to Magnitogorsk, beyond the Ural mountains—or even farther. It breaks every postulate David propounds. The copy is all in reverse type. That's a no-no which ad critic Freddie Messner pointed out when he reviewed the ad for *Printers Ink* in 1962. To make matters worse, the copy is set ragged *left* and *right* so that it literally sways back and forth to follow the wispy cigarette smoke. But in this case I agree with Messner. In 99 out of 100 cases, he and I take a dim view of body copy in reverse type, but the sacrifice in legibility in this ad is more than made up for by the unity of composition. That's really what makes the ad unusual and eye-appealing. That wisp of cigarette smoke with the headline, "A Vibro-Energy Mill can grind materials finer than particles in this smoke," and the undulating copy-block all combine to make this ad a real stopper. Bob Kenagy, Sweco's manager of advertising and public relations, gives credit for the ad preparation to Charles Bowes Advertising of Los Angeles. You have to give credit to Sweco for having the guts to break some of the rules—like those against reverse type and irregular settings—and coming up with this really great business paper ad.

Our only real competition.

Swingline makes more staplers, for more people, for more jobs than anyone else in the business.

No other stapler is built like Swingline. Or works like Swingline. Or looks like Swingline. Or sells like Swingline. In fact, we sell more staplers than any other company in America.

Swingline. The only stapler line you need to buy.

Office Products Division
32-00 Skillman Avenue, Long Island City, N.Y. 11101

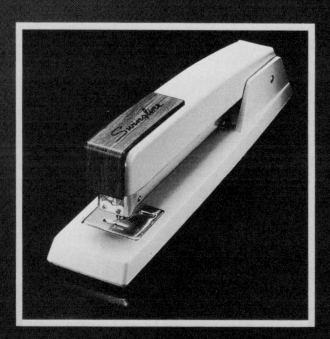

Swingline

The Swingline Company has had the lion's share of the stapler market for many years. It's an enviable position, but then there is always a chance of an erosion taking place. What do you do to remain number one and keep the competition at bay? Swingline is smart enough to know that good advertising can at least help wage a holding action. If you're the leader, you've got to act like a leader. And that's what fine advertising can do. "Our only real competition" says it all. If anyone misses the point, the first line of copy very quickly and boldly puts the reader's mind at rest: "Swingline makes more staplers for more people for more jobs than anyone else in the business." This great ad, prepared for Swingline by copywriter Jerry McTigue and art director Fred Greenfield of Al Paul Lefton Company in New York, has been running in the office products trade magazines, and it's just brilliant in its simplicity. Bravo.

3M

The 3M Company does everything right. Anyone who would change its name from the Minnesota Mining and Manufacturing Company to just plain 3M deserves a hearty vote of confidence from its many publics. And whoever is in charge of selecting photographs for this wonderful corporate series deserves a medal. My favorite in 3M's current series is:

"The toughest job here is being carried out by the deck." You better believe that tremendous action photograph wasn't just an ordinary studio shot. BBDO's creative director William Monaghan, based in Minneapolis, deserves a kiss and a kudo from Karl Kaufman, 3M's advertising manager, for this outstanding "3M Hears You" ad campaign. *Industrial Marketing's* Copy Chasers liked it so much they voted it as 1981s first runner-up in their Best Indus-

trial Advertising roundup. Frankly it should have been number 1, but then everybody's fallible. The first paragraph reads: "A carrier deck is pounded by angry seas and 60-knot winds for hundreds of thousands of miles. It must withstand the hellish blasts of the powerful, steam-driven catapults. It may take up to 200 screeching landings a day." What an intro! Action words like *pounded, angry seas, hellish blasts, powerful,* and *steam-driven* are designed to pull you into the rest of the copy or perhaps give you a severe case of mal de mer—whatever, it's darn good writing and leads right into how you can use Scotch-Brite brand Clean 'n Strip Nonwoven Abrasive to strip away oxide scale and rust. 3M claims the ad has been rated first in readership in every issue it has appeared. That's some record.

Timken

"How to" headlines always seem to pull well in business-to-business advertising. I guess that's why copywriter Terry Scullin of BBDO in New York used this proven device in conjunction with an eye-catching four-color photograph. Here again, the incongruous situation works well. Timken tapered roller bearings really do cut down a product's weight and any endomorph who has spent a lifetime dieting will readily attest to the fact that Rye Crisp on a bed of lettuce immediately signifies "how to

HOW TO LOSE WEIGHT FAST.

One of the best ways to cut down on energy consumption is to cut down on the weight of your product.

And one of the best ways to cut down on some of that unwanted weight is to design in Timken® tapered roller bearings.

Timken bearings are precision-engineered to do big jobs in little places. They can take up much less space than most other types of bearings. That means whatever you're designing can be designed smaller and lighter. Sometimes significantly smaller and lighter.

Weight reduction is one way Timken bearings can help make your product more energy-efficient. But it's not the only way.

Tapered roller bearings are, after all, just about unbeatable at beating friction—the prime consumer of energy.

If your product could stand to lose some excess poundage, we recommend a steady diet of Timken bearings. Why not start today—at the design stage?

The Timken Company, Canton, Ohio 44706.

The company that tapered the bearing can taper your bearing costs.

lose weight fast." Art director Steve Lovitch could have gone overboard and insisted on using silver flatware, a pepper mill, a salt shaker and cottage cheese in the photo, but it would have loused it up royally. Thanks for your restraint, Steve, the ad is just fine as it is. It sure doesn't look like an industrial bearing ad, and you can easily understand why it virtually jumped off the pages of *Iron Age, Industry Week* and the industrial edition of *Business Week* to consistently achieve top-of-the-book readership scores. Timken is one of the highest scoring ad campaigns ever to appear in *Business Week*.

NYLOK NEWS

AG 48

HOW NYLOK SELF-LOCKING THREADED FASTENERS SOLVE ASSEMBLY PROBLEMS...PROVIDE LASTING RELIABILITY.

NYLOK SCREWS KEEP BIG WHEELS IN PLACE

Oversized lawns are cut down to size by this heavy-duty mower. And its over-sized wheels are kept in <u>their</u> place by small Nylok screws.

Each wheel carries an oversized 3" tread tire and helps support up to 80 pounds of mower weight. Yet a single Nylok 5/16 - 18 x 1" round head standard machine screw and slip washer holds each wheel so surely that vibration and shock of heavy-duty mowing on rough lawns can't shake it loose.

The Nylok fastener was chosen because it starts easily, requires no other locking auxiliary, and stays locked against wheel rotation.

100 YR. OLD DESIGN PROBLEM SOLVED

After years of user complaints, a well-known arms maker stopped screws from loosening during repeated firing by using up to six Nylok screws per hand gun. The Nylok screws can be reused time after time but retain their torque.

Customer complaints and the considerable expense of handling, servicing, and correspondence formerly due to the loose screws problem have been eliminated. Nylok fasteners have proved a big step forward in building dealer good will and user satisfaction.

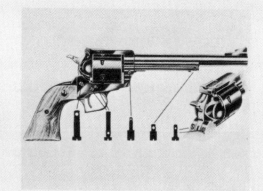

NYLOK FASTENERS KEEP SEAT BELTS SECURE

The strongest webbing materials and sturdiest buckles offer no more passenger protection than a foxtail on an antenna if the seat belt isn't reliably anchored to the car itself.

That's why practically every safety-conscious auto manufacturer uses a Nylok fastener to secure seat belts to the car floor.

Under special vibration tests producing thousands of high-G shocks per minute and equalling more than two years of hard driving, ordinary fasteners loosened in seconds but Nylok bolts stayed tight. And they do not depend upon critical torquing or special locking devices that could easily be overlooked or misapplied.

United Shoe Machinery—Nylok

Anything new, newsy or just hot-off-the-press gets readership. That's why so many ad writers resort to newsletters-in-print for advertising campaigns. You may think they're overworked and you may be right. But when you have a product that lends itself to the announcement of many new applications, a newsletter may well be the answer. In order to keep them different and truly creative, you've got to be informative as well as newsy. It's good to tell readers something they don't know, but it's even better when that something you're telling them will help them in their jobs.

And that's exactly what this United Shoe Machinery Corporation fastener ad does. Called the "Nylock News," the ad is a double-sided insert printed in a typewriter face on heavy, light blue stock. The heavy stock makes the insert stick out in a magazine and this also helps give added readership. The insert has six or seven photographs on the two pages, along with a copy block describing each cut. The reverse or backside contains a coupon for inquiries. Every copyblock carries a provocative, zesty headline that preaches to the OEM designer/reader about new fastener applications. When the ads started to gain momentum back in 1966, the competition was fierce. Industrial fasteners were produced by more than 400 companies, and USM wanted to get a much bigger share of the $1 billion market.

Proof of the company's success was the fact that soon after it was launched the campaign won a total of 14 awards from five different business publications. In *Purchasing Magazine* it won 6 of 8 possible awards the publication gave out for top Starch readership ratings attained in one issue! Our hats off to Stan Andrews, the USM ad manager, and his agency copy-contact, Paul McConville of McCann/ITSM.

Keep It Simple

Strike three.
Get your hand off my knee.
You're overdrawn.
Your horse won.
Yes.
No.
You have the account.
Walk.
Don't walk.
Mother's dead.
Basic events
require simple language.
Idiosyncratically euphuistic
eccentricities are the
promulgators of
triturable obfuscation.
What did you do last night?
Enter into a meaningful
romantic involvement
or
fall in love?
What did you have for
breakfast this morning?
The upper part of a hog's
hind leg with two oval
bodies encased in a shell
laid by a female bird
or
ham and eggs?
David Belasco, the great
American theatrical producer,
once said, "If you can't
write your idea on the
back of my calling
card,
you don't have a clear idea."

A message as published in the *Wall Street Journal*
by United Technologies Corporation, Hartford, Connecticut 06101

United Technologies

At one of the 4A's annual meetings in Boca Raton, David Ogilvy made a swan-song speech in which he pushed for advertising that gets down to the basics. He pleaded for future literate creative-types who would combine a sense of humor and a respect for research. And that's what this United Technologies ad campaign is really all about.

Since 1979, UTC has published a monthly series of full-page messages in the *Wall Street Journal*—unlike any advertising the *Journal* or its readers have ever seen. To date, they have received 253,210 requests for nearly 1,203,400 reprints!

How much more basic can you get when you write a headline like "Keep It Simple."; The copy tag line (which was a David Belasco quote), "If you can't write your idea on the back of my calling card, you don't have a clear idea," started a steady flow of business cards with ideas on their backs to United Technologies chief executive Harry Gray, whose name was hidden in 6-point type at the bottom of the ad. Many people phoned in asking for annual reports; some said they would look into buying UTC stock. Lots of people sent in resumes saying they'd like to come work for UTC. Harry Gray has received lots of gifts: books, flowers, paintings, records and tapes. The response has been incredible.

Richard Kerr, the writer of this exceptional business-to-business campaign, has no peer when it comes to getting back to basics and keeping it simple. Much of the credit also goes to Gordon Bowman, UTC's director of corporate creative programs, and his designer William Wondriska.

ADVERTISEMENT—*This entire page is a paid advertisement.*   *Prepared Monthly by U. S. Industrial Chemicals, Inc.*

U.S.I. CHEMICAL NEWS

May ★ A Monthly Series for Chemists and Executives of the Solvents and Chemical Consuming Industries ★ 1940

Shrink Fits With Solox and "Dry-Ice" Utilized on Motors

End Risk of Aero-Engine Failure From Loose Valve Seat Rings

Greatly improved fits for valve seat rings on airplane engines, with less risk of engine failure, are reported to result from chilling the parts with a solution of "Dry-Ice"* and Solox. The rings are machined, hardened and ground with a tolerance of .004", and then placed in a bath of the Solox and "Dry-Ice,"

Bushings in these locomotive parts are inserted by using Solox and "Dry-Ice" to chill and contract the bushings below the fit size.

which chills them until they are reduced below the fit size. The cylinder block receives a steam bath to expand the holes slightly, and the rings are inserted in the block.

When rings and block regain normal temperature—

*Manufactured and supplied by Pure Carbonic, Incorporated, an associated Company of U.S.I.

(Continued on next page)

Betters Paint Quality by Upping PbO:PbSO₄ Ratio

LAKEWOOD, Ohio — More durable paint films result when basic lead is introduced into drying oil vehicle paints in the form of basic lead sulphates having a PbO:PbSO₄ ratio of at least 2:1, it is claimed in a patent granted to an inventor here.

The improved quality of the paint film is said to arise from the formation of lead-oil soaps, which increase consistency and aid in forming a film with greater elasticity, lower permeability to water, and reduced rate of destructive oxidation.

Finds Actinic Rays Speed Organic Acid Formation

MIDLAND, Mich.—That the synthesis of organic acids from primary alcohols or aldehydes in the presence of a catalyst can be accelerated by actinic rays is claimed in a patent granted to an inventor here.

According to the inventor, experiments in the formation of acetic acid from a mixture of ethyl alcohol and water showed that irradiation with actinic rays gave a slightly higher yield with a catalytic mass only one-half to one-fourth as large as was needed when the actinic rays were not employed.

Ethyl Alcohol is produced by U.S.I.

Improved Dehydration Processes Open New Fields for Castor Oil

Product Offers Interesting Possibilities as Replacement For Tung Oil in Typical Paint and Varnish Formulations

Dehydrated castor oil, now produced by a method that gives 100% dehydration, is displaying excellent potentialities as a drying oil for use in the manufacture of quick drying paint and varnish, it is reported.

Characterized in its ordinary state by an iodine number of about 85, and hence classed as a non-drying oil, castor oil when subjected to the dehydration process develops good drying oil characteristics, which allow it to be used in replacing tung oil.

Compared with other tung oil substitutes, dehydrated castor oil appears to have definite advantages. It dries more quickly than perilla

New Name—Same Policy

In keeping with the constant broadening of U.S.I.'s scope of service to industry, SOLVENT NEWS appears for the first time this month under a new name — U. S. I. CHEMICAL NEWS. The editorial policy remains unchanged — to bring you every month news of the latest developments in the fields of solvents and chemicals, and of the industries that consume them.

Make Paints That Resist Water, Fire, and Mildew

BALTIMORE, Md.—How strongly adhesive paints that resist water, fire, and mildew can be formulated with zinc borate is revealed in a patent granted to two inventors here.

The paints, it is claimed, may be applied on iron and steel, wood, cardboard, and various types of wall board, and have marked affinity for galvanized iron.

A typical coating, it is said, has the following proportions:

Chlorinated paraffine	16.8%
Chlorinated rubber	6.0%
Tricresyl phosphate	1.2%
Pigment and filler	21.6%
Zinc Borate	14.4%
Solvent	40.0%

Suitable solvents include ethyl acetate and chlorinated hydrocarbon solvents.

A modification using Paris green is said to have excellent anti-fouling properties.

Ethyl Acetate is produced by U.S.I.

The castor plant is among the newest of drying oil sources. Plant height ranges up to 35 feet.

or linseed oil and forms a more durable and waterproof film, it is claimed. A further reported advantage of dehydrated castor oil is that it does not turn yellow with age.

Double Bonds

Until very recently, it was supposed that the dehydration process gave a high proportion of conjugated double bonds (conjugated diene structure), but recent work has shown that only about 25% of the molecules of the dehydrated ricinoleic acid contain the conju-

(Continued on next page)

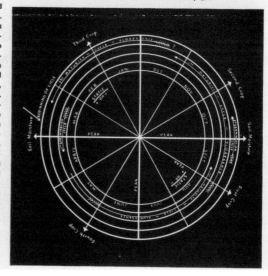

The castor plant grows both wild and cultivated in tropical regions. Until very recently the beans used in this country were all imported from Brazil. In order to provide a domestic source of supply, Woburn, Incorporated, began raising castor plants in Texas and Florida. As indicated by this chart, the castor plant is rapid-growing. Time required from sowing to harvesting is so short that three to four crops can be obtained in a single year. The castor plant displays marked variations in size, ranging in height from 6 to 35 feet. Considerable variation is also noted in the dimensions and color of the seeds. The oil from the castor beans possesses good drying oil properties after it has been dehydrated, and is currently attracting considerable interest among manufacturers of paints and varnishes as a means of replacing tung oil, which is now difficult to obtain.

U.S. Industrial Chemicals

Here;s a real old-timer that started running in 1932 as a four-page insert using the name "Solvent News." Later retitled "U.S.I. Chemical News," this informative, technical newsletter ran for over 35 years in chemical trade publications. Its newsy, editorial approach directed to chemists and executives built up a reader loyalty over the years that was absolutely fantastic. Run on blue stock, the "News" had many ancillary benefits. It was primarily an advertisement, but it also served as an excellent direct mail piece. Its terse advertorial writing gave USI's customers and prospects a veritable security blanket when reading the copy. Unlike most ads that everybody knows have been paid for, the "News" made you think it wasn't an ad because it had such an air of editorial credibility about it. According to Gerry Miknich, USI's ad director this great campaign earned fabulous readership scores over the years and was consistently rated the number one scorer in chemical publications. After the word got around, a number of publishers were deluged with calls from their other advertisers requesting positions adjoining the USI insert. Imagine that. Eat your hearts out, you Campbell-Soup-position advocates!

U.S. Steel

For the last several years, steel buyers have had a lot to worry about. They know that reliable, long-term suppliers are vital to their ongoing manufacturing operations. And smart steel buyers are also concerned about the effects of economic, regulatory and foreign trade on their potential sources of supply. The ad people at U.S. Steel decided it was an ideal time to come up with a trade ad campaign that would reassure customers that, in spite of all the country's steelmakers' problems, U.S. Steel was working its butt off to make sure future availability, quality and cost would be just what the steel user always dreamed of. The theme "The Harder You Look—The Stronger We Look" had been running since January 1979 in 17 dramatic four-color spread adver-

comparison.

The technology often comes from our own laboratories and engineering groups. For example: "stage charging" on our coke batteries, and the self-sealing oven doors, are U.S. Steel developments. Unlike many companies, we have the capability to design—and often construct—the equipment ourselves.

In spite of this progress, we're facing a maze of increasingly stringent (and often conflicting) requirements. At U.S. Steel, we frankly question whether they are all reasonable, affordable, or even make environmental sense.

Some of the huge sums involved could be used to increase production efficiency and create new jobs. We've spoken out on this issue, which is vital not just for us, but for all manufacturing industries.

However, our special capabilities, and our history of achieving environmental solutions help us make the wisest choices for the very considerable investments we put into environmental controls every year.

And this, in turn, helps assure you of continued steel supply just as much as our technology, or our reserves of raw materials. Surely the vast resources and self-reliance of U.S. Steel must play a part when you are formulating your long-range steel-buying plans.

THE HARDER YOU LOOK— THE STRONGER WE LOOK.

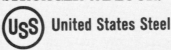 United States Steel

tisements. Directed to people who influence metalworking purchasing policy, they're still appearing in *Iron Age, Industry Week, Purchasing, Purchasing World, Metal Progress, Production* and *Machine Design*. Their readership is fantastic. John Wallace, U.S. Steel's ad manager of the manufacturing divisions, claims that 44 of these ads have been measured and that 52% of them placed first among materials ads in the issue—86% in the top two. This particular ad provided the highest readership in *Iron Age* for the entire year of 1979. And don't forget that's the top ad of 36 consecutive yearly issues. And it also was the top ad in the metalworking category during all of 1980. The campaign also won a prestigious American Business Press O&R award in 1980. Thatsa some record!

"Duck Out of this Dogfight...
GUN'S DOWN TO 13 ROUNDS!"

He who fights... *and can tell when to run away*... may live to fight another day. So when a plane's machine gun tells the gunner how many rounds are left after each "line of typewriting," the gunner simply tells the pilot... and the pilot hits for home.

But what tells the gunner his gun is almost a goner? A Veeder-Root Counting Device that subtracts—from a total set at the number of rounds available—the exact number of bullets in each burst of fire, *as they are fired*. One glance at the counter-face says plainly to the busy gunner: "Stay" or "Go".

All of which goes to show that Veeder-Root Devices can be "good soldiers" in time of war, just as they're good salesmen, production controllers and public servants in time of peace. In either case, the "facts-in-figures" they faithfully supply are always valuable ... often vital ... even, as here, the measure of difference between life and death. Chances are that they can profitably serve *you* ... built into your product or installed on your production equipment. May we have a chance to figure out *how*?

[On this page September 30: How Veeder-Root Counts in the Making of a Great Newspaper.]

VEEDER-ROOT
INCORPORATED
Factories at Hartford and Bristol, Connecticut
Offices in Principal Cities Throughout the World

Veeder-Root

Back in 1940, there was an awful lot of sabre-rattling and chauvinistic breast-pounding. All the smart copywriters were taking advantage of this upsurge of nationalism to get better readership. Three writers—P.M. Abbott, J.H. Squier and J.A. Keary of Boston's Sutherland-Abbott ad agency—got together and produced this fine piece of work for their client, Veeder-Root. They knew alliteration, when used sparingly, can often add to readership scores. It also helps make dull headlines come alive. In this case, when combined with an excellent headline, it came up with a real grabber: "Duck Out of this Dogfight...Gun's Down to 13 Rounds." I defy anybody not to want to read the copy. It starts out "He who fights... and can tell when to run away...may live to fight another day." What a great lead-in! Sheer poetry to the macho audience it's designed to reach. The ad tells an exciting story to executives on how Veeder-Root counters help tell various types of equipment (in this case machine guns) exactly where they stand.

The dramatic illustration of the cockpit-enclosed pilot is greatly enhanced by the spot drawing of a downed fighter trailing black smoke from its tail in the background.

To top it all off, Veeder-Root used a very smart sign-off to help insure good readership for future ads in this series. Bracketed copy reads: "On this page September 30: How Veeder-Root Counts in the Making of a Great Newspaper."

Volkswagen

Volkswagen spends megabucks on consumer advertising, but the company is smart enough to budget a lot for business publications as well. Doyle Dane Bernbach, Volkswagen's great ad agency, uses the same dynamic copy and art approach in its truck ads as it does in its consumer car advertising. They're all of the same genre and they are absolutely beautifully done. My favorite is this one headlined: "Ain't ours." Can you imagine the guts it took to use such a daring vulgarism as "ain't"—in a *headline* no less! Only Volkswagen and DDB could get away with it and make it pay off big in attracting readers. Who else would pay a lot of money to show, even in an uncomplimentary way, a competitor's product? The copy explains very fast why the truck stuck in the snow drift ain't a Volkswagen: "For one thing, our engine's in the rear. No weight sits on our drive wheels. It's like putting concrete blocks in your trunk. For another, most trucks only clear about 7 inches, but the VW clears 9½. Our bottom doesn't drag."

And neither does your copy. It's written in terse, staccato sentences that read like the writer is actually talking to the reader. Volkswagen's marketing manager, Paul R. Lee, deserves a lot of credit for letting Doyle Dane Bernbach have its head and approving the use of some colloquial language that is so easy to read. You can bet DDB doesn't do what a lot of big consumer agencies do—put the tyro, lower-salaried copy and art teams on the industrial accounts and use the pros for the consumer stuff. Good thinking.

Wang revolutionized word processing 5 years ago. And today everyone else is five years behind.

Five years ago, we revolutionized word processing by introducing the first word processor that was even easier to use than a typewriter. With a TV-like screen that displayed what you were typing.

So if you wanted to correct mistakes, add or delete words, or rearrange entire paragraphs, you simply did it all on the screen. In seconds. When you wanted to print finished copy out, a separate printer hummed along at 200 words per minute. And you could store up to 120 pages of information on a small diskette.

People have been so pleased with our word processing in fact, that Wang is #1 in the world today.

Over 50,000 companies are using our systems to save time and money. Bankers in Hong Kong. Camera-makers in Munich. And 85% of the Fortune 500. That kind of experience has helped us understand the keys to increasing productivity in almost any kind of business. It's also helped us develop the only word processing system that lets you grow from a single workstation to a system supporting up to 128 workstations, depending upon application. With word processing, data processing and electronic mail capabilities.

And that's what's kept us five years ahead of everyone else.

1976. Our first ad.

Wang Office Information System.

For a demonstration of Wang word processing, call

1-800-225-0643

(in Massachusetts call 1-617-459-5000, extension 5711). Or send this coupon to: Wang Laboratories, Inc., Business Executive Center, One Industrial Ave., Lowell, MA 01851

Name _____
Title _____
Company _____
Address _____
City _____ State _____ Zip _____
Telephone ( ) _____

Making the world more productive.

Wang Laboratories

Here's an ad within an ad. In this ad, Wang Laboratories chose to show a picture of the 1976 ad it ran to introduce the world's first word processor. It very plainly emphasizes this with the headline "Wang revolutionized word processing 5 years ago. And today everyone else is five years behind." A fantastic concept that really worked. One-half of the ad space is taken up by the body text, but it's set in clear, big 18-point type for easy reading. Jack Wallwork and Dick Pantano, the copy/art team at Hill, Holliday, Connors, Cosmopulos (Wang's agency) must have invented the body-copy typeface because I couldn't match it in any type book. Suffice to say, though, it reads, and it reads beautifully. That in itself makes the ad as unusual as it is outstanding.

Who says black (and white) isn't beautiful? Wang didn't have to pay three or four times the premium it costs to run four-color advertising to get results. According to Matt Sperber, the ad guy at Wang, this little gem which requires a "hard lead" (a lead for a demonstration is the *only* response) produced several hundred demonstrations and generated one million dollars in revenues. Now that's some response!

Wonder what a Frenchman thinks about

Two years ago a Frenchman was as free as you are. Today what does he think—

—as he humbly steps into the gutter to let his conquerors swagger past,

—as he works 53 hours a week for 30 hours' pay,

—as he sees all trade unions outlawed and all the "rights" for which he sacrificed his country trampled by his foreign masters,

—as he sees his wife go hungry and his children face a lifetime of serfdom.

What does that Frenchman—soldier, workman, politician or business man—think today? Probably it's something like this—"I wish I had been less greedy for myself and more anxious for my country; I wish I had realized you can't beat off a determined invader by a quarreling, disunited people at home; I wish I had been willing to give in on some of my rights to other Frenchmen instead of giving up all of them to a foreigner; I wish I had realized other Frenchmen had rights, too; I wish I had known that patriotism is *work*, not talk, *giving*, not getting."

And if that Frenchman could read our newspapers today, showing pressure groups each demanding things be done for them instead of for our country, wouldn't he say to American business men, politicians, soldiers and workmen —"If you knew the horrible penalty your action is bound to bring, you'd bury your differences now before they bury you; you'd work for your country as you never worked before, and wait for your private ambitions until your country is safe. Look at me... I worked too little and too late."

WARNER & SWASEY
Turret Lathes
Cleveland

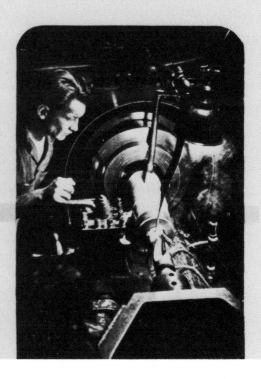

YOU CAN TURN IT BETTER, FASTER, FOR LESS... WITH A WARNER & SWASEY

1941

Warner & Swasey

My test mailing to the Business/Professional Advertising Association brought in many appeals from members to feature the Warner & Swasey free enterprise series that was prepared by its agency, Griswold-Eshleman. The wartime ad "Wonder What a Frenchman Thinks About" was produced in 1941 and was designed to convey to Warner & Swasey's many publics that "this was the kind of a company you would want to do business with." Requests for reprints came by the barrel. They totaled—get this—a whopping 4,400,000. They were picked up in newspaper editorials, used in newspaper advertisements *paid for by others,* utilized by radio commentators (we didn't have many TV sets in those days), stuffed into pay envelopes, used as plant posters, appeared in house organs, folders, you name it! The copy tugs at your heart and espouses the kind of chauvinism we used to identify with years before the feminist movement stole the word and recoined it to use as its own battle standard. In the ad, the subjugated Frenchman thought, "if you knew the horrible penalty your action is bound to bring, you'd bury your differences now before they bury you; you'd work for your country as you never worked before, and wait for your private ambitions until your country is safe. Look at me... I worked too little and too late." Sound familiar? The ad could run today and be just as meaningful and thought-provoking as it was 40 years ago.

Westinghouse

A few years ago in an *Advertising Age* article, I panned a four-color Westinghouse spread because it used as a main illustration a lot of light bulbs formed into a map of the United States. Right after that, all hell broke loose!

I got two letters from *Business Week*, bawling me out because that ad attained BW's top-of-the-book readership. I got a letter from Fred Harvey, the agency's account management supervisor, who was an old friend and wanted to become an ex-friend. I got another terse letter direct from Westinghouse telling me what a great job the ad had been doing making Westinghouse "a powerful part of people's lives." Then to top it all off, six months later my agency was bought out by Westinghouse's agency! Believe me, I'm still walking a tightrope. Now I have a chance to rectify my gaffe. A couple of B/PAAers have suggested I

Westinghouse technology goes with it.

tout this 1981 Westinghouse ad which tells a very convincing story about how Westinghouse technologies helped build, fly and defend the F-16 multimission fighter. Tom Van Steenbergh, the D'Arcy-MacManus & Masius art director, gets all the credit for laying out a very dramatic aerospace ad. His accomplice, Joe Dayton, wrote the no-nonsense copy which, by "telling-it-as-it-is," received a big 24% Read Most Starch score in *Business Week*. That was twice the four-color spread norm. Bob Lukovics, manager of corporate advertising at Westinghouse, should be proud of this one (it beats the old monochromatic map-and-light-bulb combination I got my fanny in a sling for castigating). Starch data tells us that "wherever an F-16 climbs..." was the highest scoring four-color spread measured in any advertiser category in *Business Week* in *10 years*.

Wheeling Steel

Any economist worth his salt could list the problems of the economy in the mid 1960s and ditto them for the early 1980s. (It's a sure way to understand the real meaning of déjà vu.) Inflation in 1966 was running rampant. Bank interest rates had hit an all-time high, and the government made banks put the kibosh on borrowing. While this was happening, Wheeling Steel, the country's tenth largest producer, was slipping fast. It was sorely in need of better, more economical production facilities and had allocated $175 million for capital expenditures. Deliveries, product quality and customer service were truly the pits. A new advertising-minded president by the name of Robert M. Morris came in and saved the day. In order to give the company a new image and bring competition to its heels, he invented an exclusive thing called P.T.O.—Price at Time of Order. Up until then, if a customer ordered steel, the

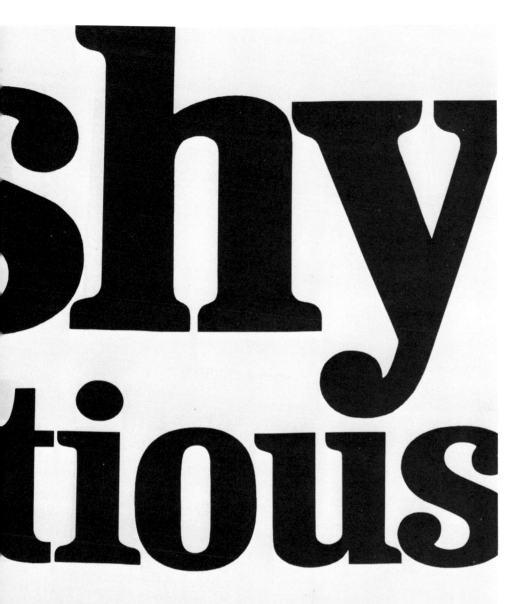

price was quoted at time of delivery. Morris changed all that by guaranteeing prices for six months at the time of the order.

His agency, Gardner Advertising, came up with this classic all-type "Hustle" theme which was created by copywriter Jack Silverman with art direction by Bob Reed. Best of the bunch of hustle ads is this "Pushy Ambitious" one which is representative of the whole campaign. The huge all-lowercase headline is a *real* headline. It's reminiscent of what the front page of many daily newspapers do to sell more papers. Face it, banner headlines work—especially when you don't expect to see them in a trade publication. The aggressive copy is as competitive and forceful as the strategy behind the campaign. And it worked. According to Jan Avrigean, Wheeling's director of marketing services, and *Industrial Marketing*, which named it number one campaign of 1966, this fabulous Wheeling campaign was an industrial Avis. That's for sure.

HAULING JEEPS ACROSS RAVINES

Another Job for Wickwire Rope

There are many *extra* jobs that wire rope must do today. The Army, the Navy, and the Liberty Fleet need vast quantities of wire rope.

This points up the message: MAKE YOUR WIRE ROPE LAST LONGER.

When wire rope fails too soon, it wastes badly needed steel, labor and time. If by care you can make *your* wire rope last longer, you are helping to win the war.

When you *do* need wire rope, be sure to get the correct type and size for your service. Our representatives will help you determine the best Wickwire Rope, preformed or Wissco lay, for your use — and after installation will help you get longest life and most economical operation from it.

The reason for the long life of *Wickwire* rope lies in our controlled production, every step of the way from ore pile, to blast furnace, to open hearth, to finished rope.

* * * *

Have you a copy of the book "Know Your Ropes"? It will help you get longer rope life. More than 25,000 wire rope users all over the world regard it as the authoritative guide to wire rope economy. For your free copy, write Wickwire Spencer Steel Company, 500 Fifth Avenue, New York, N. Y.

WHEN WIRE ROPE LEAVES ITS IMPRINT in sheave or drum groove there is a filing action on the rope at every start and stop. Grooves having the slightest impressions should be machined and polished. This and forty more rope life-savers are pictured and described in "Know Your Ropes."

SEND YOUR WIRE ROPE QUESTIONS TO WICKWIRE SPENCER

 **WICKWIRE ROPE**

Sales Offices and Warehouses: Worcester, New York, Chicago, Buffalo, San Francisco, Los Angeles, Tulsa, Chattanooga, Houston, Abilene, Texas, Seattle. Export Sales Department: New York City

WISSCO PRODUCTS

Wickwire Spencer Steel

The "Big War," commonly known as World War II, created a lot of shortage. Stepped-up war demand for essential military products switched industry from the manufacturing of butter to the more vital production of guns. During that time, there was a bunch of trade advertisements devoted to reminding people that war production was of paramount concern. Copy platforms of the era took great pains to encourage readers to make use of the materials they had and to take proper care of them in order to obtain maximum product life. Ad after ad would spell out the importance of proper maintenance.

This fine Wickwire Rope effort set the style for these wartime ads. Written in 1942, this old veteran used a superb illustration of a soldier-therefore-hero (a surefire way to get readership) sitting in a jeep suspended from two pulleys. The pulleys were riding on a single strand of Wickwire rope stretched over a deep valley. The headline, "Hauling Jeeps Across Ravines—Another Job For Wickwire Rope," tells it exactly as it is, in a glance. The stunning artwork has just enough depth and focus to give the page-thumber acrophobia as he turns from one page to another. The reader may not want to walk close to the edge of this precipice, but he has to be drawn to the page by his own curiosity and daring. Once he's sucked into the body copy by the main illustration and headline, the text tells him how to make wire rope last long. And if that isn't enough to help with the war effort, the last paragraph insists you write for a copy of the booklet, *Know Your Ropes*. All this of course is bound to help the Allies achieve longer rope life (and hopefully win the war). It must have done the job because, as the ads point out, more than 25,000 users of wire rope already had copies of the book. And we did win World War II.

Xerox has come up with another miracle.

"Astounding"..."Amazing"..."A miracle."
That's what the experts said when they saw our new XL-10 Imaging Process.

A dramatic breakthrough in copy quality.

It actually gives you offset-quality copies. Blacker blacks and whiter whites than you've ever seen before.

Even the tiniest details leap out crisp and clear.

Of course, a copy is worth a thousand words. So if you send your business card to us at Xerox Square, Rochester, New York 14644, or call 800-648-5600, operator 260,* we'll send you back an XL-10 copy. Or arrange a demonstration.

Once you see it, we think you'll agree with our experts.

XL-10 copies aren't just positively beautiful. They're absolutely faithful.

The XL-10 Imaging Process.

Xerox

Living black and white can produce readers (and miracles) when it's done right. Brother Dominic, the beneficent Franciscan monk of television fame, has done a splendid job of selling and maintaining Xerox's position as *the* leader in the copier field. Needham, Harper & Steers, the Xerox agency, was smart enough to take advantage of exploiting Allen Kays' Clio-Hall-of-Fame 1975 TV commercial and adapting it for use as a black and white space campaign. Here's a fine example of what can happen when you team up electronic media with print media. They both prosper and serve as a valuable adjunct to one another. This

spread was part of a four-page bleed unit that ran in *Fortune*. It tied for first place as Best Read Business Advertising in 1980. This award is given annually to the business advertiser who attains the highest average Recall Reading score for all *Fortune* ads measured by Harvey Research. No small accomplishment to say the least, because most *Fortune* advertisers are tough competitors insofar as garnering high readership scores is concerned. It could be that Needham, Harper & Steers has started a new fad. From now on it may be de rigueur to use dogs, babies, horses *and monks* in ads to attract readers. Incidentally Brother Dominic happens to be a retreaded Jewish comedian who launched his career on the borscht circuit in the Catskills. How's that for typecasting?

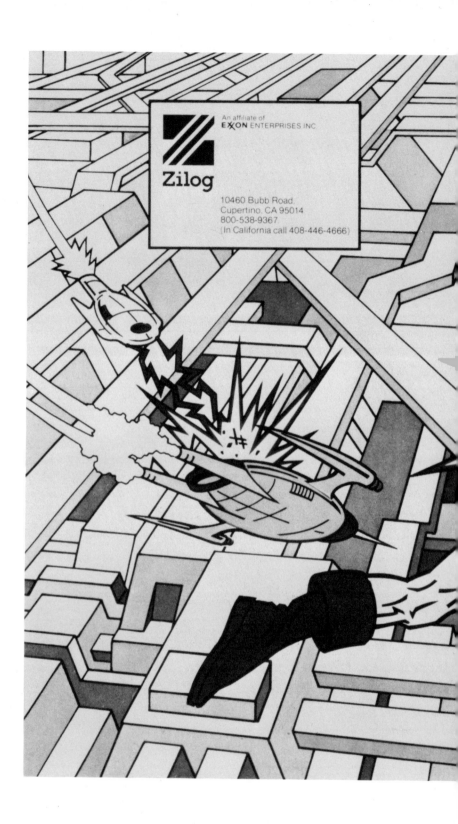

Zilog

Zilog, an affiliate of Exxon, wanted a bunch of leads for its Z-8000 microcomputer. The company also wanted to dominate the very important Wescon trade show and the special show issue of *Electronic Design Magazine* (which is always packed with hundreds of attractive ads that are vying for the subscriber's attention). Bob Pinne, creative director of Zilog's West Coast agency, Pinne, Garvin & Hock, came up with an idea that guaranteed domination and readership in EDN's thickest issue of the year. He prepared a 20-page media spectacular that combined the weight and heft of an insert with the surefire pulling power of a full-color comic

book. And to make sure it brought in inquiries (and could also be measured), he added two stapled-in business reply cards which offered a free 18 x 24 color poster of Captain Zilog, the protagonist/hero of the comic book. An ancillary benefit to be derived here was to build a ready prospect list for Zilog and its distributors. There was no way any reader thumbing through that 1979 issue of EDN could pass by this humorous but informative insert. Just its sheer size and bulk alone assured Zilog of the magazine falling open to its ad. Written and illustrated by comic book designer Lou Brooks, the advertisement brought in over 5,000 inquiries within two weeks after it ran. Now that's pulling power, Captain!

Bibliography

Alcoa:	*Industrial Marketing.*	January 1939.
Andersen Corporation:	*Industrial Marketing.*	January 1976.
Armco:	*Industrial Marketing.*	September 1936.
Carborundum:	*Industrial Marketing.*	January 1966.
Celanese:	Dartnell. *Dartnell's Advertising Managers Handbook.*	1969. page 710.
Clark Equipment:	*Magazine Age.*	December 1981.
Container Corporation of America:	*Advertising Age.*	
Durkee:	*AdWeek*	May 1981.
Hercules:	*Industrial Marketing.*	April 1939.
International Paper:	*AdWeek.*	
ITT:	Flesch, Rudolf. *The Art of Readable Writing.* New York: Harper & Brothers	
Motorola:	*AdWeek.*	
Pitney-Bowes:	*Industrial Marketing.*	January 1967.
Rockwell:	*Industrial Marketing.*	January 1960.
Rome Wire:	*Class & Industrial Marketing.*	March 1927.
Sweco:	*Printers Ink.*	June 1962.
3M:	*Industrial Marketing.*	January 1982.
Veeder-Root:	*Industrial Marketing.*	October 1940.
Volkswagen:	*Industrial Marketing.*	January 1960.
Warner & Swasey:	Watkins, Julian Lewis. *The 100 Greatest Advertisements:* Who Wrote Them and What They Did. Dover Press	
Wheeling Steel:	*Industrial Marketing.*	January 1967.
Wickwire Spencer Steel:	*Building Magazine.*	September 1981.

Index

Index

Abbott, P.M., 185
Akers, Kenneth W., 67
Allied Chemical, 38
Ally & Gargano, Inc., 53
American Cyanamid, 38
American Home Products, 113
American Rolling Mill Company, 16, 17
Anderson, Cal, 55
Andrews, Stan, 177
Artzybasheff, Boris, 84
Avrigean, Jan, 195
N.W. Ayer, Inc., 13, 42

Baker, Cindy, 83
G.M. Basford Company, 59, 64, 157
BBDO, 81, 172
Ted Bates Worldwide, Inc., 84
Belasco, David, 179
Benson, Howard, 95
Bernbach, Bill, 84
Biederman, Barry, 95
Blakeley, Jim, 130
Blaschke, Stefen, 127
Charles Bowes Advertising, 169
Bowman, Gordon, 179
Bradley, Dave, 15
Bromberg, George, 88
Brooks, Lou, 201
Broyles, Allebaugh & Davis, Inc., 62
Budd, Monty, 71
Bugbee, Alan, 72
Business/Professional Advertising
  Association, 45, 62, 105, 111, 151, 191, 192
Byck, Lew, 134

Calhoun, Don, 9
Campbell-Ewald, 147
Campbell-Mithun, Inc., 15, 99
Capawana, Rod, 33
Cecil, George, 13
Cloyd, Wade, 10
Cody, Iron Eyes, 3
Cody, Sherwin, 23
Copeland, R. Bruce, 161
Costa, Phil, 121
Crain, G.D. Award, 41
Creamer Inc., 5, 50, 140, 157
Cunningham & Walsh, 55, 159

Dana, Marshall, 123
d'Arazien, Art, 97

D'Arcy-MacManus & Masius
  Worldwide, Inc., 31, 193
Darling, Oliver, 23
Davis Advertising, 19
Dayton, Joseph, 193
Dettman, Morris, 81
de Garmo Inc., 143
de Garmo, John, 143
DeWolf, John, 21, 105
Donahue & Coe, Inc., 71
Doremus & Co., 57
Downs Crane & Hoist, 41
Doyle Dane Bernbach, Inc., 84, 117, 126, 134, 187

Eckersley, Thomas, 123
Ellington, Ken, 143
Evans, Keith J., 151

Falter, John, 84
Farago, Peter, 87
Farin, Philip, 92
Fenske, Mike, 119
Flesch, Rudolph, 95
Fletcher/Mayo Associates, Inc., 90
Foote, Cone & Belding Adv., Inc., 83
Fuess, Bill, 92
Fujita, S. Neil, 123
Fuller & Smith & Ross, Inc., 5, 10, 115

Gallup & Robinson, 133
Gardner Advertising, 195
Geer Du Bois, 87
Giusti, George, 123
Goldberg, Joe, 3, 113
Gramm, Gene, 57
Grapper, Hank, 74
Gray, Harry, 179
Greenfield, Fred, 171
Grey Conahay and Lyon, Inc., 42
Griswold-Eshleman, 67, 191
Gulbransen, Cliff, 140

Haddad, Richard, 107
Harris, Herb, 157
Harvey, Fred, 192
Hatcher, Ed, 64
Henkel Inc., 7
Hennesy, Larry, 63
Herbrecht, Al, 113
Hight, Jack, 96
Hill, Holliday, Connors, Cosmopulos, 189

Hill, Jon, 35
Hillman, Hans, 123
Hoffman, Dustin, 136
Horning, Hugh B., 43
Hovenack, Richard, 92
Huggins, John, 79
Hughes, Ed, 131
Hungerford, David, 111

Jackson, John, 87
Jacot, Pierre, 130
Jewett, Richard K., 133
Johnson, Ray, 96

Kaufman, George S., 55
Kaufman, Karl, 172
Kay, Allen, 198
Keary, J.A., 185
Kenagy, Bob, 169
Kerr, Richard, 179
Ketchum Communications Inc.
Ketchum, MacLeod & Grove, 27, 136, 137
Killelea, Richard, 89
Klaproth, David, 99
Knudsen, Bunkie, 97
Kollewe, Chuck, 126
Kopf, Fred, 42
Kravec, Joseph, 137
Kuhlman, Ray, 123

LaBeaume, E.I., 71
Lawson, Ed, 59
Lee, Paul R., 187
Al Paul Lefton, Inc., 171
Lewis & Gilman, Inc., 7
Linck, Bruce, 49
Linnehan, Leonard, 101
Lionni, Leo, 123
Lord, Geller, Federico, Einstein, Inc., 84
Loudon, Henry A., Inc., 101
Lovitch, Steve, 175
Lubalin, Herb, 123
Lukovics, Robert, 193
Lynch, Brad, 13
Lyon, George, 42

MacManus Advertising, 25
MacManus, Theodore, 25
Marsteller, Bill, 113

Marsteller, Inc., 3, 74, 110, 112, 113, 121, 145
McCann-Erickson, 9
McCann-Itsm, 177
McConville, Paul, 177
McFee, William, 17
McGraw, Hugh, 3
McTigue, Jerry, 171
Marvin, Theodore, 71
Massard, Frank, 28
Massey, John, 37
Mau, Richard, 164
Meldrum & Fewsmith, Inc., 45
Mellebrand, Elmer, 147
Messner, Fred, 169
Michel Cather, Inc., 60
Miknich, Gerry, 181
Millar, Robert F., 41
Mintz & Hoke Inc., 107
Moeller, Bob, 31
Monaghan, William, 172
Mondrian, Piet, 125
Morris, Gil, 115
Morris, Robert M., 194
Morrison, Dick, 123
Moser & Cotins, 149

Nathan-Garamond, Jacques, 123
Needham, Harper & Steers Adv., Inc., 95, 198, 199
Novak, Gene, 28

O'Daly, Fergus, 79
Ogilvy, David, 37, 169, 179
Ogilvy & Mather, Inc., 92, 152, 153
O'Sullivan, Dalton, 55

Paepke, Walter P., 37
T.N. Palmer & Company, 96
Pantano, Dick, 189
Paonessa, Joseph, 147
Pasch, Robert, 152, 153

Pataky, Pat, 64
Patterson, Jeff, 131
Pawluk, Hal, 162
Perry, Chris, 45
Pinne, Garvin & Hock, Inc., 130, 200
Pinne, Robert, 130, 200
Price, Roger, 101
Price, Ted, 109
Putnam, Kent, 19

Rand, Paul, 123
Reagan, Nancy, 13, 95
Reagan, Ronald, 65
Redford, Robert, 13
Redmond, Rich, 110
Reed, Bob, 195
Reeves, Rosser, 84
Richardson, Norman, 101
Rike, George, 134
Ritz, Bob, 99
Robertson, Jim, 10
Rockwell, Norman, 84, 139
Rockwell, W.F., Jr., 145
Roth, Larry, 103
Rozinski, Ted, 50
Rumrill Hoyt, Inc., 28

Santi, Aldo, 123
Saxon, Charles, 47
Scali, McCabe & Sloves, 164
Scherm, Al, 10
Schmertz, Herbert, 117
Scott, J.I., 35
Scott, Jim, 35
Scullin, Terry, 174
Sekiguchi, Ken, 157
Serkowich, Joe, 103
Shaw, Allan, 60
Shortlidge, Tom, 119
Silldorf, Henry, 59
Silverman, Jack, 195
Singer, Bob, 74, 110, 121

Slesar, Henry, 115
Smith, Stan, 64
Spectrum Marketing, 23
Sperber, Matt, 189
Squier, J.H., 185
Stanger, Ron, 45
Stevens, Van, 140
Struthers, George H., 152, 153
Sutherland-Abbott, 185
Sutnar, Ladislov, 123

Taubin, Bill, 126
Tesch, Michael, 53
J. Walter Thompson, 47
Thuna, Leah, 153
Tinker, Jack, 9
Todd, Ken, 60
Truman, Harry, 151
O.S. Tyson & Company, 123

Union Carbide, 38

Van Patten, H.E., 67
Van Steenbergh, Tom, 57, 193
Venti, Sal, 140

Wallace, John, 183
Wallwork, Jack, 189
Warner, Bicking & Fenwick, 33
Warner, Jack, 33
Warren, Bob, 15
Watts, Ralph, 147
Weir, Walter, vii
Welch, Raquel, 10
West, Weir & Bartel, 31
Whittey, Rick, 107
Williams, John, 141
Wolf, Henry, 123
Wood, Dean, 50
Worcester, Dale, 5

Young, Scott, 137
Young & Rubicam, Inc., 113, 118